D0034725

ECOLOGICAL ANTHROPOLOGY

ECOLOGICAL ANTHROPOLOGY

Donald L. Hardesty
University of Nevada, Reno

Alfred A. Knopf

New York

First Edition

98765

 All inquiries should be addressed to Newbery Award Records, Inc., 201 East 50th Street, New York, N.Y. 10022. Published in the United States by Newbery Award Records, Inc., a subsidiary of Random House, Inc., New York and simultaneously in Canada by Random House of Canada Limited, Toronto. Distributed by Random House, Inc.

Library of Congress Cataloging in Publication Data:

Hardesty, Donald L. 1941-
Ecological Anthropology

Bibliography: p.
Includes index.
1. Human ecology. 2. Anthropology. I. Title.
GF41.H38 301.31 76-25424
ISBN 0-394-34407-3

Manufactured in the United States of America

To Susan and My Parents

PREFACE

I have two reasons for writing this book. First, I am convinced that an ecological perspective can do much to help students of anthropology find the unifying threads in their discipline. Students have often complained to me about their difficulty in understanding how specialized courses in simple and complex societies, past and present societies, biological and cultural man, and linguistics have enough in common to be part of a single field of study. Since I teach a wide variety of courses in a small university, the complaint has also crossed my mind. Furthermore the question of "relevance" always lurks in the background, demanding an explanation of the ways in which anthropological studies can help cope with contemporary problems. The ecological approach, with its emphasis upon interrelationships, seems admirably suited to answering these questions.

The second reason is simply that I have been unable to find a textbook that meets the demands of a course in ecological anthropology or that can be used for supplementary reading in courses where an ecological approach is used. Julian Steward's *Theory of Culture Change* (1955), once useful, no longer covers the field as it is currently defined. There is a variety of books consisting of case studies in "cultural ecology" and at least one book of readings, but none is intended to give a general background. Finally, the one or two modules that could be used as a general introduction are not comprehensive enough to be useful as a textbook.

There are some real problems in writing a book of this kind. Many people pointed out to me that ecological anthropology is not yet a unified science; indeed, some feel that it is no more than a "point of view." Furthermore, those who do see the approach as unified are not always in agreement about its underlying assumptions and goals. While I recognize that ecological anthropology is in an early stage of development there are numerous unifying concepts and principles. These have formed the foundation of this book.

The book is planned for advanced undergraduate students and graduate students in anthropology. Nevertheless, much of the material is useful in more elementary courses; a glossary has been included to make the book more usable at that level. The glossary will hopefully also make the book suitable for students of human ecology who are not specializing in anthropology, particularly those in the biological sciences. Finally, I must emphasize that the book is not intended to outline human history from an ecological perspective. Rather, it focuses upon basic concepts and principles.

The book is organized into three parts. Part 1 considers the concept of the ecological system. It is intended to give the student an overview of the ecological relationships into which humans enter. Part 2 takes another look at ecological relationships but this time from the viewpoint of the human ecological population. Finally, Part 3 covers two special topics in ecological anthropology. Ethnoecology, the first, focuses upon the environment as perceived by humans rather than as viewed by an outside observer. Since I did not feel that I could do complete justice to the approach, Professor Catherine S. Fowler, of the University of Nevada, Reno, kindly consented to write this chapter. The second topic, human paleoecology, discusses how fossil data can be used to define and to understand ecological relationships in the past.

It is impossible to acknowledge the sources of all the ideas that went into the preparation of this book. However, I particularly want to give credit to Andrew P. Vayda for influencing my way of thinking. All or part of the manuscript was read by Peter Comanor, Don Dumond, William Douglass, Kenneth Knudson, Stuart Marks, Betty Meggers, Allan Swedlund, and Andrew Vayda. The manuscript benefited greatly from their comments. In addition, I want to acknowledge the useful comments of several anonymous reviewers and the assistance of my editor Herbert J. Addison. Finally, thanks are due to Barbara Taylor who typed most of the manuscript, often several times, with patience and diligence. Without these people the book would not have been possible.

Donald L. Hardesty

CONTENTS

ONE
INTRODUCTION

The science of anthropology has traditionally been a "holistic" discipline. Anthropologists have advocated a broad, comparative study of human behavior in the search for general laws and principles, and little about man has been left out. It is perhaps not surprising, then, to find that anthropological "explanation" has also been far-ranging in its attempts to make order out of the chaos of human diversity. At one time or another anthropologists have explained human behavior with reference to current topics in biology, ecology, history, evolution, diffusion, and independent invention, for example. The purpose of this book is to explore the ways in which "environment" is used in anthropological explanation, an area of endeavor currently referred to as *ecological anthropology*. The roots of ecological anthropology are to be found in several different traditions of environmental explanation, some of which are tightly woven into Western thought. Let us begin by examining these roots.

ENVIRONMENTAL DETERMINISM

Perhaps the most pervasive theme is the belief that the physical environment plays the role of "prime mover" in human affairs. Personality, morality, politics and government, religion, material culture, biology—all of these and more have at one time or another been subject to explanation by *environmental determinism*. The *humour theory* of Hippocrates was probably the single, most important foundation for environmental determinism until the nineteenth century. (This discussion of humour theory is based on Glacken, 1967, pp. 80–115.) Hippocrates saw the human body as housing four kinds of "humours"—yellow bile, black bile, phlegm, and blood, representing fire, earth, water, and blood, respectively. The relative proportions of the four humours caused variation in individual physique and personality, as well as in sickness and health. Climate was believed to be responsible for the "balance" of the humours and, therefore, for geographic differences in physical form and personality. Thus people living in hot climates were passionate, given to violence, lazy, short-lived, light, and agile because of an excess of hot air and lack of water.

The effect of climate on personality and intelligence determined other human affairs, particularly government and religion. Both Plato and Aristotle associated climate with government, viewing temperate Greece as the ideal climate for democratic government and for producing people fit to rule others. Despotic governments, on the other hand, were best suited for hot climates because the people lacked spirit and a love for liberty and were given to passionate excesses. Cold climates had no real form of government because the people lacked skills and intelligence and were strongly given to a love of individual liberty.

The eighteenth century Frenchman Montesquieu continued this line of reasoning and applied it to religion. Hot climates create lethargy, according to this scholar, and are apt to be associated with passive religions. Buddhism in India was given as a classic example. By contrast, religions in cold climates, Montesquieu believed, are dominated by aggressiveness to match the love of individual liberty and activity. (Christianity, Montesquieu's religion, was elevated above environmental determinism because it was *revealed*.) The geographer Ellsworth Huntington (1945) carried this thinking well into the twentieth century by arguing, in the *Mainsprings of Civilization*, that the highest forms of religion are found in temperate regions of the world. His basic argument was that temperate climates are more conducive to intellectual thinking.

The nineteenth and early twentieth centuries brought a decline in the popularity of humour theory but no less vigorous apologists for environmental determinism. There are several reasons for its persistence. The developing method of science was marked by the search for simple, linear, cause-and-effect relationships; that is, *A* causes *B* causes *C*, and so forth. There was no recognition of the complex interactions and feedback processes that make today's science. Anthropologists and geographers searched for simple causes of the geographical distribution of culture traits. Some proposed environment while others favored diffusion. Both offered simple, straightforward explanations that were consistent with linear science. Therefore, it is not surprising to see the resurgence of environmental determinism at this time. The rise of "technological determinism," as espoused by Marxist social philosophy, also contributed to the resurgence. Environmental determinism was a rebuttal to the antienvironmental position of Marxist writers. Finally, an explanatory model of this kind was a simple way to categorize and explain the mass of data on human diversity being accumulated as a result of world exploration, in much the same way that the "Three-Age System" helped classify ancient artifacts. The "culture area" concept was particularly suitable for this purpose, allowing diverse cultures within large geographical areas to be classified into a single type because some traits are held in common. Some early geographers and anthropologists quickly noted the general correspondence between culture areas and natural areas and argued that environment *caused* the occurrence of distinct cultural areas.

Material culture and technology were believed to be most affected by the environment. For example, in a discussion of the prehistory of the American Southwest, William H. Holmes, a turn-of-the-century anthropologist, states that

it is here made manifest that it is not so much the capabilities and cultural heritage of the particular stock of people that determines the form of material culture as it is their local environment. (1919, p. 47)

However, nonmaterial culture was also explained environmentally. F. W. Hodge, editor of the *Handbook of American Indians North of Mexico,* published in 1907, explains, also with regard to the American Southwest, that

the effects of this environment, where the finding of springs was the chief desideratum in the struggle for existence, were to influence social structure and functions, manners and customs, esthetic products and motives, lore and symbolism, and, most of all, creed and cult, which were conditioned by the unending, ever-recurring longing for water. (p. 430)

Perhaps the most lucid proponent of environmental explanation for nonmaterial culture was J. W. Fewkes, another turn-of-the-century American anthropologist, who was particularly interested in the origins of ritual behavior. However, unlike most of his contemporaries, Fewkes was aware of the *complexities* underlying the study of man-environmental interaction and did not assume a simple one-to-one relationship. (e.g., 1896, p. 699)

Today the theme of environmental determinism has been largely replaced by the emergence of man-environmental models that assign environment a "limiting" but uncreative role or that recognize complex mutual interaction. However, the explanation of biological diversity in humans continues to have a strong, deterministic orientation. Models of genetic change in human populations, for instance, are still dominated by the theory of natural selection, a theory that assigns to environment a strong and active role in shaping gene pools. Thus the most popular explanation for the distribution of skin color is based upon "selection" for pigment gradules that help block out excessive ultraviolet radiation from the sun. Models of physiological adaptation to altitude and temperature are also marked by environmental determinism. On the other hand, a number of recent investigators have suggested models that greatly limit the role of environment as an agent of biological change. "Genetic drift," that is, vagaries due to sampling errors in small populations, is an important part of most of these models. The role of natural selection is particularly being questioned because of the recognition that genes are not isolated entities subject to easy manipulation by environmental factors but are part of complex systems of interaction.

POSSIBILISM

The general orientation of environmental explanation in anthropology shifted away from determinism and toward *possibilism* in the 1920s and 1930s. Much of this shift was due to the personal influence of Franz Boas who showed that the origin of *specific* cultural features and patterns was generally to be found in historical tradition rather than in environment. Boas's emphasis upon specific cultural explanation gave rise to the so-called school of "historical particularism," a school that has often been chided for its antienvironmentalism. However, Boas did not completely ignore environment: "I shall always continue to consider . . . (environmental variables) . . . as relevant in limiting and modifying existing cultures" (M. Harris, 1968, p. 266). He did consider it irrelevant to explaining the *origin* of culture traits. Environment, then, played an important role in explaining why some features of culture did *not* occur but not in explaining why they *did* occur. This belief is the hallmark of possibilism.

Perhaps the most famous example of possibilistic explanation is that posited by Alfred L. Kroeber (1939) for the geographical distribution of maize cultivation. Kroeber gave data showing that the distribution of maize farming in aboriginal North America was restricted to climates with at least a four-month growing season, during which time rainfall was sufficient and there were no killing frosts. A similar study was made by the archaeologist Waldo Wedel (1941) who proposed that on the aboriginal Great Plains the geographical boundaries of farming was a function of rainfall. Farming was practiced only in areas where the mean annual rainfall was sufficiently high to assure the necessary growing season and in areas where drought was not frequent. In areas where the mean annual rainfall was high enough but in which killing droughts were frequent, mixed farming and foraging (hunting and gathering) were practiced. Finally, in areas with both frequent droughts and low average rainfall, only foragers were found.

Possibilism made significant contribution to the "culture area" concept. As early as 1896, Otis T. Mason suggested that the geographical distribution of material culture and technology is "molded" by the environment but is not caused by it (1896, p. 663). He defined 12 "ethnic" environments or culture areas based upon this assumption. Mason's work was elaborated by Clark Wissler (1926) and A. L. Kroeber (1939). Both recognized a general correlation between cultural areas and natural areas but viewed the correlation in terms of what culture features a natural area would or would not permit. Thus farming was diagnostic of the eastern United States, not because the temperate climate caused it but because it permitted the necessary growing season. Likewise, big game hunting was permitted by the grasslands of the Great Plains, after the introduction of the horse and firearms, but was not caused by it. Finally, the limited cultural development in the Great Basin and other "marginal" areas was attributed to environmental limitations while the cultural "florescence" in the southeast United States was attributed to the *absence* of environmental limitations. Environment, however, could not be used to explain why one culture area was marked by patrilineal inheritance and another by matrilineal inheritance. This could only be explained by culture history. Thus Kroeber remarked that

> *while it is true that cultures are rooted in nature, and can therefore never be completely understood except with reference to that piece of nature in which they occur, they are no more produced by that nature than a plant is produced or caused by the soil in which it is rooted.* The immediate causes of cultural phenomena are other cultural phenomena [*italics mine*]. *(1939, p. 1)*

The culture area concept, therefore, developed into a kind of compromise between determinism and the extreme diffusionist views of the "kulturkreis" and related schools.

The role of environment in cultural evolution is particularly clear in possibilist thought: Environment places stringent limitations upon the level of cultural development. Perhaps the most frequently cited example of this position is that taken by the archaeologist Betty Meggers. In her 1954 paper "Environmental Limitations on the Development of Culture," Meggers suggests that farming is necessary for advanced stages of cultural evolution and that an area's suitability for farming is an accurate measure of its "potential" for cultural evolution. She defines four environment "types," from least suitable for farming to the most suitable (1972, pp. 179–80):

Type 1. "Where agriculture is impossible because temperature, aridity, soil composition, altitude, topography, latitude, or some other natural factor inhibits the growth or maturation of domesticated plants."

Type 2. "Where agricultural productivity is limited to a relatively low level by climate factors causing rapid depletion of soil fertility."

Type 3. "Where relatively high crop yields can be obtained indefinitely from the same plot of land with fertilization, fallowing, crop rotation, and other kinds of soil restorative measures, or in more arid regions by irrigation."

Type 4. "Where little or no specialized knowledge is required to achieve and maintain a stable level of agricultural productivity."

These types are not to be construed as *causing* cultural evolution. According to Meggers, Types 3 and 4 may not reach a high level of development for *cultural* reasons, for example, the absence of appropriate diffusion. However, no amount of diffusion or other cultural factors can lead to advanced cultural development in a Type 1 or Type 2 environment (1954, p. 822). Furthermore, if an advanced culture expands into a Type 1 or Type 2 environment, it is doomed to failure.

The most notable application of this model is in the lowland Maya region. Meggers has argued for a long time that the lowland Maya, who occupied a Type 2 environment, migrated into the tropical lowlands *after* achieving the roots of civilization elsewhere, probably in the highlands of southern Mexico and Guatemala (e.g., 1972, p. 65). Reaching maturity in its new home, it suddenly surpassed the farming potential of the poor environment and collapsed.

THE ECOLOGICAL PERSPECTIVE

Environmental determinism and possibilism have one thing in common: an *Aristotelian* view of the relationship between man and environment. That is, humans occupy one sphere and environment another and never the twain shall meet. The purpose of both models is to ascertain the impact of one sphere on the other, with the deterministic view holding that environment actively shapes man (and vice versa) and the possibilistic view assigning environment a limiting or selecting role. According to the anthropologist Clifford Geertz,

with such a formulation, one can ask only the grossest of questions: "How far is culture influenced by environment?" "How far is the environment modified by the activities of man?" And can give only the grossest of answers: "To a degree, but not completely." (1963, p. 3)

A more precise understanding of the relationship between man and environment is made possible by the non-Aristotelian view that constant *interplay* takes place and that two distinct "spheres" do not exist. One cannot be understood without the other. This assumption provides the theoretical foundation of *ecology*, the third major theme in environmental thought. The word *ecology* was apparently first coined by the German biologist Ernst Haeckel to refer to the way that animals make their living, "above all the beneficial and inimical relations with other animals and plants" but also including relationships with the inorganic environment (as quoted in M. Bates, 1953). However, the roots of the ecological orientation are found deep in Western tradition. Indeed, the idea of interplay occurs in the writings of Plato and Aristotle and later in the literature of the Judaeo-Christian tradition. Thus Aristotle introduced the concept of "design in nature," which viewed the earth and the universe as a clock-like system having interrelated parts, although parts that were distinct "species" and driven by different causes. The concept of design in nature was picked up by various Judaeo-Christian philosophers and elaborated to suit their purposes, including the substitution of the Aristotelian "final cause" with the Christian diety as the reason for the design. Clarence Glacken, a geographer, refers to this school of thought as *physico-theology* and gives the following passage to suggest its prevailing view of the world:

the earth (is) an orderly, well-planned place in which there was "nothing wanting, nothing redundant or frivolous, nothing botching or ill-made". (1967, 422)

Physico-theology took on a more secular orientation with its inclusion into the "normal" science of the eighteenth and nineteenth centuries. The earth and the cosmos were viewed as perfect clockworklike mechanisms subject to completely predictable *natural laws* rather than divine purpose.

Continuity of the ecological theme into the 19th century is particularly expressed in Charles Darwin's "web of life" concept and in the writings of Baron Alexander von Humboldt (Glacken, 1967, pp. 375–428). According to Darwin all living creatures must mutually adjust to each other in their "struggle for existence." In the *Origin of Species* (1859) he gives an example of the web of life, as he refers to this relationship. Bumble bees are responsible for pollinating clover fields in the English countryside. However, the abundance of the bees is limited by field mice because they destroy the hives. With fewer bees less clover is pollinated and the field is not as productive as it might be. Darwin observed, however, that clover fields were more productive near villages and towns. Why? House cats were abundant in these localities and preyed upon field mice, thus drastically reducing their numbers. With fewer field mice bumble bees boomed and fields blossomed.

Humboldt, an early 19th century German naturalist and traveler, had views similar to Darwin. He was particularly interested in the relationship between plants and man in the tropical regions of the world. According to Humboldt man is often responsible for changing the character of native plants by introducing exotics that become dominant, driving the indigenous species out of existence or into remote places. The most typical result was the creation of *monotomy* of landscape, eliminating natural diversity in favor of a few plants useful to man. However, plants have a corresponding impact upon man. Humboldt believed that plant diversity, such as that in the tropics, stimulated *human imagination* and *artistic sensitivity*. (He traveled widely in the American tropics and was quite impressed by the ruins of Precolumbian civilizations in the jungles. It is not surprising that the spectacular remains led him to this conclusion.) When man replaced the natural diversity with monotomy, the quest for knowledge and artistic endeavor suffered accordingly.

Cultural Ecology

Ecology as a science blossomed in the twentieth century but has been mostly restricted to the study of plants and animals other than man. However, the ecological vantage point in anthropology was expressed as early as the 1930s by Julian H. Steward. Perhaps the most important contribution of his "method of cultural ecology" was the recognition that environment and culture are not separate spheres but are involved in "dialectic interplay . . . or what is called *feedback* or *reciprocal* casuality" (Kaplan and Manners, 1972, p. 79). Two ideas essential to the ecological viewpoint are inherent in the concept of reciprocal causality: the idea that neither environment nor culture is a "given" but that each is defined in terms of the other, and the idea that environment plays an *active*, not just a limiting or selective, role in human affairs. At the same time it must be kept in mind that the relative influence of environment and culture in a feedback relationship is not equal (Kaplan and Manners, 1972, p. 79). According to this view, sometimes culture plays a more active role and sometimes environment has the upper hand. Steward believed that some sectors of culture are more prone to a strong environmental relationship than other sectors and that ecological analysis could be used to explain cross-cultural similarities *only* in this "culture core." The culture core consisted of the *economic* sector of society, those features that are "most closely related to subsistence activities and economic arrangements" (Steward, 1955, p. 37). The "method" of cultural ecology then involved the analysis of

1. The interrelationship between environment and exploitative or productive technology.

2. The interrelationship between "behavior" patterns and exploitative technology.

3. The extent to which those "behavior" patterns affect other sectors of culture (Steward, 1955, pp. 40–41).

Steward's culture core did not include many aspects of social structure and almost no ritual behavior. Neither of these were considered to be significantly related to environment. Furthermore, Steward excluded from cultural ecology the study of biology, stating that "culture, rather than genetic potential for adaptation, accommodation, and survival, explains the nature of human societies" (1955, p. 32).

Cultural ecology retained the possibilist's interest in the study of *specific* cultural features. Steward's goal was "to explain the origin of particular cultural features and patterns which characterize different areas" (1955, p. 36). His method required that *detailed* studies of local groups in their environment be conducted as a prerequisite to making ecological generalizations (Vayda and Rappaport, 1968, p. 489). There can be no question that this focus is responsible for the present florescence of ecological studies in anthropology.

Andrew Vayda and Roy Rappaport (1968, pp. 483–87), while recognizing the importance of Steward's contribution, have criticized his approach as being inadequate. Steward gives as his primary objective the explanation of *origins* of certain cultural traits. However, his approach is to show first how a cultural feature and an environmental feature *covary*, that is, how they are functionally interrelated, and second, to show that the same relationship *recurs* in historically distinct areas. Vayda and Rappaport argue that this approach does not necessarily mean that the environmental feature *caused* the cultural feature for the following reasons:

1. Sampling procedures are not adequate to eliminate the possibility of spurious correlations.

2. Even if the correlations are statistically significant, correlations do not necessarily mean a cause and effect relationship.

3. Even if significant correlations and causality were shown, it does not necessarily mean that the relationship is inevitable, as Steward believed.

The second weakness of Steward's cultural ecology was to treat the culture core as if it included *only* technology. Several studies, as we shall see, have shown that ritual and ideology also interact with the environment. Vayda and Rappaport point out additionally that Steward's selection of environmental features for study does not include other organisms (e.g., disease microorganisms) nor does it include other human groups, perhaps the greatest weakness of all. (However, more recent applications of cultural ecology have taken "social environment" into consideration, and fruitfully so.) Finally, his approach does not include the study of interaction between culture and biology, neither genetic nor physiological. Yet numerous studies have shown that culture and biology go hand in hand in several areas, such as nutrition, and that one cannot be understood without the other.

Population Ecology

The study of the environmental relations of *particular* human groups, introduced by Steward's cultural ecology, marked the beginning of population ecology in anthropology. The *ecological population* is a local group of organisms, belonging to the same species, with a distinctive life style; that is, the members of the group get food in the same, distinctive way, have essentially the same tolerances for things in the environment, are fed upon by the same predators, and so forth. Population ecology is the study of those processes that affect the *distribution* and *abundance* of ecological populations. External processes affect a population's relationships with food, water, weather, and other organisms, among other things. By contrast internal processes include such things as behavioral, physiological, and genetic responses to population density. The study of populations has several distinct advantages for human ecology. In the first place the population is a more or less "bounded" unit, subject to quantitative description and analysis. Populations can be counted, their size and distribution measured, and so forth. In the second place the ecological population is the traditional unit of study in nonhuman ecology. Therefore, human populations

> *are commensurate with the other units with which they interact to form food webs, biotic communities, and ecosystems. Their capture of energy from, and exchanges of materials with these other units can be measured and then described in quantitative terms. No such advantage of commensurability obtains if cultures are made the units, for cultures, unlike human populations, are not fed upon by predators, limited by food supplies, or debilitated by disease. (Vayda and Rappaport, 1968, p. 494)*

In other words the concept of the ecological population gives a quantifiable common denominator suitable for the study of human and nonhuman ecology alike.

The suggestion by Vayda and Rappaport that the population be made the *focal point* of human ecological studies should not be taken to mean that anthropologists have not until now been interested in population ecology. Julian Steward's cultural ecological studies in the 1930s on the Great Basin and Southwestern Indians are explicitly aimed at understanding the relationship between human populations and their environment. Steward particularly studied the effect of environment upon the *distribution* of human populations. Thus in his classic study of the aboriginal Great Basin, he focused upon "interrelated physical features of the landscape" as determinants of population distribution, including water, altitude, temperature, geographical barriers, and the annual variation in the occurrence of eatable foods (1938). The intellectual descendants of Steward have continued his interest in the distribution of populations through *settlement pattern* studies. A settlement pattern is the disposition of a population's settlements vis-à-vis the natural landscape and the "social" environment. The determinants of a settlement pattern are numerous but include such things as physical barriers, technology and subsistence, political organization, kinship, warfare, and ideology and symbolism (Trigger, 1968). Complex interaction among these variables is responsible for the actual distribution of a population. Settlement pattern studies have particularly been stressed in archaeology, receiving an important impetus from Gordon Willey's (1953) study of the Viru Valley in coastal Peru. Since the publication of that study in 1953, interest in the interrelationship between the distribution of a population and its physical and social environment has been continued. Most recently, that interrelationship has been formalized into an explicit "systems" approach, such as the "settlement-subsistence system" of Stuart Streuver (1968b).

Steward also was interested in the interrelationships between environment and the *abundance* of human populations. However, his orientation in this regard was possibilistic. Thus, in the Great Basin, he saw population density as "correlated with the fertility of the natural environment" (1938, p. 48). The possibilistic view of population abundance was also expressed by A. L. Kroeber in his classic work *Cultural and Natural Areas of Native North America,* published in 1939. Kroeber drew upon earlier population estimates of North American tribes and his own estimates to support a general correlation between "natural" areas and population density. He concluded that "other things being equal, we (can) infer a denser population from a richer ecology, or, among agriculturalists, from a larger area of more fertile soil" (1939, p. 180).

In 1953 Joseph Birdsell published a landmark study of the relationship between *mean annual rainfall* and population abundance in aboriginal Australia. Rather than define environment in terms of "fertility" or "ecological richness," Birdsell chose a single variable, mean annual rainfall, to represent the relevant environment. Birdsell's reasoning was that mean annual rainfall determines plant growth and that plant growth sets limits on the food available to humans, either directly through plant foods or indirectly through animal foods. Consequently, variation in mean annual rainfall (allowing for "unearned" water from, for example, rivers with drainage basins outside the area under study) should be correlated with population density, and Birdsell's study supported such a correlation. Note that the study is possibilistic in orientation, since increased mean annual rainfall *allows,* but does not cause, higher population densities. The concept of *carrying capacity* has been used in similar studies (e.g., Zubrow, 1975). Carrying capacity is the theoretical limit to which a population can grow and still be supported permanently by the environment. As a population approaches this limit, it places "pressure" on the environment to provide the resources needed for subsistence. Population pressure, in turn, sets into motion demographic forces to limit further growth, forces that include *cultural* means of reducing birth rates (abortion, contraception) and increasing death rate (warfare).

The carrying capacity of an environment depends, of course, upon the subsistence methods available to a population and can be changed by technological innovations making possible more efficient exploitation or production. The late archaeologist V. Gordon Childe, in fact, saw population growth as being *dependent* upon subsistence (e.g., 1951). Thus the abundance of foragers is severely restricted by a low carrying capacity, but the adoption of farming raises the carrying capacity and makes possible a "population explosion." Childe coined the name "neolithic revolution" to stress the relationship between farming and population growth, among other things.

In recent years the "possibilistic" view of population growth has been questioned by a number of persons. Many anthropologists now accept a *causal* model of population growth, and a popular, if controversial, position is that taken by the economist Ester Boserup. Boserup (1965) argues that population growth causes a higher carrying capacity by forcing people to use land more intensively and to adopt technological innovations that make more intensive land use possible. For example, a shift from an extensive method of farming, such as slash-and-burn, to a more intensive method based upon fertilization and irrigation is brought about by population pressure, and that shift increases the carrying capacity of the land. (Observe that carrying capacity is not conceptualized as a constant but varies according to technology and environment.) Others have taken a less deterministic stance and see population growth and subsistence as *interdependent*. That is, population pressure stimulates technological innovations that, in turn, not only make more intensive land use possible but also increase the carrying capacity and stimulate a new population explosion. This kind of mutual causality is essential to ecological thinking, as we have noted.

The study of human population ecology is plagued by the same problem of definition that has handicapped so many other endeavors. The boundaries of local groups are not always clear, often blending into neighboring groups and rendering arbitrary the definition of the population. Perhaps more important is that local human groups are not always economically independent and, therefore, do not constitute a distinctive life style. For example, local groups of advanced farmers may produce a few specialized crops and depend upon trade with other groups to obtain whatever else is needed. The "ecological population" in this case is not the local group but *all* of the groups participating in the trading network. A similar problem is encountered with the concept of the genetic or "Mendelian" population. The genetic population is defined by a mating system, that is, a group within which most mating takes place. However, local groups of humans are not always synonomous with a mating system. Marriages may be arranged with persons in neighboring groups or groups farther away. In these cases the effective genetic population is the *network* of local or more distant groups, a network that is sometimes called a "commune" or "connubium." The problem of definition is obviously critical to any kind of quantitative or comparative study and will be discussed throughout this book.

Systems Ecology

The publication of Clifford Geertz; *Agricultural Involution* (1963) was another milestone in ecological anthropology. His approach is rooted in cultural ecology, indeed that is what he calls it, but his perspective is based upon the concept of the *system.* A system is a "set of objects together with relationships between the objects and between their attributes" (Hall and Fagan, 1956, p. 18). Instead of focusing upon "reciprocal causality" between two objects or processes, the system focuses upon a *complex* network of mutual causality. The methods of systems analysis are used, first, to define the boundaries and environment of a system and, second, to model its complexity in such a way that system behavior can be studied and predicted.

Geertz believes that the concept of the *ecosystem* (we shall use *ecological system* to avoid the biological bias associated with the name ecosystem) is the logical conclusion to the idea of constant interplay between culture, biology, and environment. Theoretically, the ecological system is a dynamic set of relationships between living and nonliving things through which energy flows and materials cycle and because of which other problems of survival are worked out. In practice the ecological system is identified by a group of plants and animals, along with their nonliving environment, that make up a "food web" and generally affect each other's chances of survival.

The ecological system in which humans participate can be studied, according to Geertz, in the same general way as those in which humans do not. This method of study

> *is of a sort which trains attention on the pervasive properties of systems* qua *systems (system structure, system equilibrium, system change)* rather than on the point-to-point relationship between paired variables of the "culture" and "nature" variety [*italics added*]. *The guiding question shifts from: "Do habitat conditions . . . cause culture or do they merely limit it?" to such incisive queries as: "Given an ecosystem defined through the parallel discrimination of cultural core and relevant environments, how is it organized?" "What are the mechanisms which regulate its functioning?" "What degree and type of stability*

does it have?" "What is its characteristic line of development and decline?" "How does it compare in these matters with other such systems?" And so on. . . . (1963, p. 10)

Although Geertz presents an elegant argument for the use of systems theory, he does not take advantage of the conceptual and analytical tools that are associated with the theory. In effect, he uses it only as a point of view. Some later studies, however, have done so, particularly those of Richard Lee (1969, 1972a, 1972b) and Roy Rappaport (1968) Rappaport's study, for example, is concerned with energy relationships between the Tsembaga Maring (New Guinea) farmers and the ecological system in which they participate. Carefully quantified data on caloric and protein consumption, physiological stress, energy expended in subsistence activities, carrying capacity, limiting factors, and demography were collected. Among other things, he was able to show that the ritual killing of domestic pigs by the Tsembaga Maring

helps to maintain an undegraded environment, limits fighting to frequencies that do not endanger the existence of regional populations, adjusts man-land ratios, facilitates trade, distributes local surpluses of pig in the form of pork throughout the regional population, and assures people of high-quality protein when they most need it. (1968, p. 224)

Rappaport's work demonstrates dramatically the advantages of using the concepts and tools of systems theory.

That such "holistic" studies in ecological anthropology are still in their infancy, however, is suggested by the criticism that Rappaport's work has drawn (J. Anderson, 1973, pp. 199–200):

1. The geographical scale of the ecological system used to study the Tsembaga Maring is too small to understand many relevant ecological processes (e.g., material cycles and other limiting factors),

2. His data on human nutrition are taken from a sample that is too small and representative of only a short time period, and their interpretation is too simplistic,

3. The analysis of quantitative data is not sufficiently sophisticated (only descriptive mathematics were used), and his use of system theory is incomplete,

4. He does not extend the study of energy flow to include exchanges among human groups.

The systems approach to ecological anthropology has also been criticized for emphasizing the processes of self-regulation and neglecting "disruptive" processes that upset systems and cause evolutionary change (e.g., E. Anderson, 1972; Diener, 1974). That is, the ecological systems in which humans participate are viewed as delicately balanced "machines" kept in equilibrium by mechanisms that counteract deviations. In large part this view has come from ecologists studying "natural" systems; however, the recent literature strongly suggests that nature is not in a state of balance but subject to "traumas and shocks imposed by climatic changes and other geophysical processes" (Holling and Goldberg, 1971). The ecologist Crawford Holling (1973) points out that the "self-regulation" of ecological systems really includes two types of processes: equilibrium and resilience. Equilibrium processes act to keep a system from *fluctuating* too much and may not be that important in nature. By contrast, resilience processes act to prevent the system from self-destructing, to make sure that it *persists* through time. These are the important processes and under some circumstances may actually give an advantage to systems that fluctuate rather than to those that are stable. For example, if energy flows in "spurts" through a system, a species or population that is able to grow rapidly and then to "crash" to a much smaller size without becoming extinct has a selective advantage. Vayda and McCay (1975) suggest that studies of the ecological systems in which humans participate should shift from equilibrium to resilience processes. It should also be pointed out that some anthropologists using a system approach have studied instability and evolutionary change, most notably the archaeologist Kent Flannery (e.g., 1972b). Nevertheless, the early stage of development of systems ecology in anthropology is apparent.

Ethnoecology

All of the approaches to ecological anthropology that have been discussed to this point are "objective;" that is, they study man-environmental relationships from the *observer's* point of view. Another approach that has gained popularity in recent years attempts to study ecological relationships from the *participant's* point of view. Usually called ethnoscience, ethnographic semantics, or cognitive ethnography, it uses the concepts of structural linguistics to get at what Rappaport (1963) calls the "cognized" environment, the environment that is actually perceived by a human group (J. Anderson, 1973, p. 188). Informants are used to construct "folk classifications" of nature with the assumption that the classifications are clues to the way that people have coped with environmental problems. Rather than being a first step toward understanding the processes of human ecology, however, folk classifications are all too often an end in themselves. A detailed discussion of the ethnoecological approach is found in Chapter 12.

SUMMARY

The study of the relationship between humans and their environment has been of interest to scholars for a long time. In anthropology environment has been used to explain cultural origins and diversity in at least three ways: environmental determinism, environmental possibilism, and ecology. In recent years the ecological approach has replaced the other two and today is one of the most popular kinds of anthropological explanation. Nevertheless, ecological anthropology should not be viewed as the "anthropology of the future." Problems in anthropology are varied and, as in any other science, require a variety of explanations. Nor is it likely that an ecological approach will provide the best answers to *all* problems with which it can potentially deal. Human behavior is simply too complex to be understood by a single set of principles. At the same time ecological anthropology has been shown to provide a powerful explanation for some questions, questions that will be discussed throughout this book.

PART ONE ECOLOGICAL SYSTEMS

TWO
ADAPTATION AND EVOLUTION

It is axiomatic that all forms of life must somehow eat, protect themselves from hazards, and reproduce. Humans, and possibly some other organisms, also have psychological "wants" that must be satisfied to a greater or lesser extent. All of these problems are environmental in the sense that their solution is, at least in part, beyond the organism itself. In its most general sense, *ecology* is the study of the interactions between an organism and its environment that provide those solutions. *Adaptation* is the central concept in ecological studies because it is the process whereby beneficial organism/environment relationships are established. Since the explanation of how adaptation takes place is the dominant problem in the modern theory of *evolution,* ecological and evolutionary approaches cannot easily be separated from each other. According to Mayr (1970, p. 6) evolutionary theory explains adaptation as a two-stage process: (1) the production of individual varieties; and (2) the "sorting out" of the variants by natural selection. When applied to genetically inherited biological variants, natural selection is no more than differential reproduction, increasing the proportion of individuals in future generations having "adaptive" characteristics while decreasing the proportion of those that don't. By definition, then, individuals having "beneficial" characteristics reproduce, on the average, more than others. Adaptation is thereby equated with, and measured by, reproductive success. *Darwinian Fitness* is an index of reproductive sucess that is often used to measure individual or group adaptation. It is the average number of offspring produced by a variant divided by the average number of offspring produced by the variant having the greatest reproductive success. For example, if an individual with variant A produced 2.0 offspring per generation and variant B, the individual with the greatest reproductive success, produced 4.0 offspring, the Darwinian Fitness *(w)* of variant A is 2.0/4.0 or .50. Because of the method of calculation the highest Darwinian Fitness is always 1 and the lowest 0. Darwinian Fitness can also be converted to a *selection coefficient*(s) that measures how effectively variants are being "sorted" out, given by $s = 1 - w$. The selection coefficient runs from 0, the constant value for the variant with the greatest reproductive success, to 1, for individuals producing no offspring at all.

The "reproductive" concept of adaptation has the advantage of being subject to quantification and, therefore, to mathematical rigor. However, it should also be clear that, used in this way, adaptation means no more than the relative *survival* of individual variants from one generation to the next. Presumably reproductive survival is a measure of how well individuals have ultimately coped with the basic problems of life and is necessary to understanding *evolutionary* changes. Nevertheless, many of the environmental relationships established by an organism over a short time period have nothing to do with reproductive survival but with *wellbeing*. This is especially true of humans who often respond to environmental phenomena to satisfy psychological wants only occasionally related to survival. The study of human ecology, the understanding of how humans cope with their environment must, therefore, use a less restrictive concept of adaptation. A number of biologists and social scientists have suggested that the concept be expanded to include *any* beneficial response to the environment, only one of which is reproductive success (e.g., Hamburg, Coelho, and Adams, 1974; Slobodkin, 1968; Mazess, 1975; R. Thomas, 1975). Distinct measures of adaptation must be used for each kind of beneficial adjustment. Thus, in discussing human biological responses to high altitudes Mazess (1975) has proposed several different "domains" or adaptive regions in which adaptation takes place. The assessment of adaptation in each domain "involves the evaluation of (a) relative benefit or (b) degree of necessity, relative to an environmental stress" (1975 p. 170). In individuals, for example, the domains are physical performance, nervous system functioning, growth and development, nutrition, reproduction, health, cross-tolerance and resistance, affective functioning, and intellectual ability. Adaptation to each domain is measured in different ways; for example, affective functioning is measured by tests of happiness and sexuality.

There are additional problems with the concept of adaptation that must be mentioned before going further. First of all, in any integrated system, such as an organism, a change in one part requires compensation in other parts. No change is independent; all change has ramifications throughout the system. This suggests that there are really two kinds of adaptation: *external* and *internal* (Alland, 1970). External adaptation is the process of making beneficial adjustments to the environment, while internal adaptation is the process of beneficial compensation for those adjustments *within* the organism or other system. The study of adaptation must consider both kinds of processes. Second, because there are many different kinds of environmental problems and many different responses to each, adaptation is a *compromise* process. Responses that are beneficial to different problems may conflict with each other and reduce their effectiveness. It is, therefore, not surprising to find that adaptive changes often do not seem to go as far as they logically should. The problems of internal adaptation must be solved before further change can take place.

LEVELS OF ADAPTATION

The process of adaptation takes place at three "levels": behavior, physiology, and genetic/demographic (Bateson, 1963; Slobodkin, 1968; Slobodkin and Rapoport, 1974). Each level includes several adaptive domains. Rapid adjustments to sudden changes in the environment take place at the behavioral level. For example, an organism flees from a predator, moves into the shade to get out of the hot sun, or builds a fire to keep warm in a cold climate. Behavioral responses are rapid and particularly suitable for temporary fluctuations in the environment. If the disturbance continues, however, physiological, or second-level, mechanisms replace or support those at the behavioral level. For example, if, after building a fire, the organism is still cold, shivering is initiated, metabolic rate increases, and peripheral blood vessels constrict in order to conserve heat loss. Physiological responses are not as fast as behavioral responses, nor are they as varied. Finally, long-term and permanent changes in the environment are combated with mechanisms at the very heart of the organism: genetic, or third-level, mechanisms. Changes in genetic structure require several generations and, consequently, are not desirable if changes in the environment are merely temporary. Activation of the three "response" levels is both direct and hierarchical. At the time of the environmental shock, *all* levels are activated simultaneously to reduce the impact of the stress. However, the *rate* of activation is hierarchical; the behavioral level is activated most rapidly, followed by the physiological level, and finally by the genetic level.

Behavioral Adaptation

Behavior is the most rapid response that an organism can make and, if based upon learning rather than genetic inheritance, is also the most flexible. Two kinds of behavior may be adaptive. *Idiosyncratic* behavior includes all of the *unique* ways in which individuals may cope with environmental problems and tends to be the subject matter of psychology. Anthropologists, however, focus on the adaptive responses made by individuals and groups by means of *cultural* behavior. Cultural behavior is patterned, shared, and traditional and is the most distinctive feature of the human species. (I am here using culture in the ethological sense, e.g., Kummer, 1971, rather than in the more specialized sense of "symbolic behavior.") For purposes of simplicity, adaptation may be viewed as taking place because of changes in three kinds of cultural behavior: *technological, organizational,* and *ideological.* Technology is the "tool kit" used by humans in the quest for food, protection, and reproduction, including everything from digging sticks to nuclear power plants. Organizational culture is the network of social statuses and roles relating individuals in a group. It includes kinship, social rank and stratification, voluntary associations, and politics, among other things. Ideology, finally, is a rubric that includes "values, norms, knowledge, themes, philosophies and religious beliefs, sentiments, ethical principles, world-views, ethos, and the like" (Kaplan and Manners, 1972, p. 112).

Technological, organizational, and ideological changes help humans adapt in at least four ways: (1) by providing basic solutions to environmental problems; (2) by improving the effectiveness of those solutions; (3) by providing adaptability; and (4) by providing awareness or recognition of environmental problems. Cultural ecology has traditionally been concerned with the first kind of adaptation. Thus Julian Steward's concept of the culture core was visualized as a basic solution to the problem of wresting food from the environment. The usefulness of this orientation has been shown in a number of studies, one of the more interesting being Oliver's (1962) study of the bison hunters of the North American plains. Hunters and gatherers, for example, the Comanche, and farmers such as the Cheynenne acquired both the horse and the gun in the seventeenth and eighteenth centuries and moved into the high plains to hunt bison. Although their different cultural traditions introduced an element of diversity, the social organization of the plains hunters is remarkably similar. Oliver argues that these similarities are fundamental adaptations to bison hunting. Group structure, authority and leadership, and status are the primary means of adaptation. The plains hunters were organized into loose tribes integrated by secret societies and by pantribal rites such as the Sun Dance. Neither clans nor any other cross-cutting kinship units were present, even though some of the tribes with a farming tradition, such as the Cheyenne and the Teton Dakota, originally had clans. The primary group was the local band and during most of the year the bands were widely scattered over the countryside. Only during the summer were the bands clustered together into the great camp circles characteristic of the tribe. Furthermore, it was only during the summer that the pan-tribal secret societies and rites were operative. Oliver contends that this type of organization is necessary for the effective hunting of bison. Bison are herding animals that are sometimes clustered into groups of thousands and hundreds of thousands. They are also migratory but, contrary to popular belief, their migrations seldom took place along regular trails from Texas in the wintertime to Canada in the summer. Rather, the herds "swayed" a few hundred miles north and south in search of better grazing and protection from the weather. The actual movements were unpredictable and extensive scouting was required to detect where the herds were from one day to the next. Also, the concentration of the herds varied during the year. During most of the year the bison were widely scattered into small groups that were best hunted by local bands. The band camps moved frequently in pursuit and never consisted of more than a few hunters, who hunted individually. Bands were composite and individuals frequently moved from one band to the next. Any kind of corporate responsibility on the part of the group was next to impossible to maintain and, for that reason, kinship units such as the clan were not suitable. However, in the late summer and early fall the dispersed bison congregated into dense masses to breed, leaving large parts of the plains completely deserted. At this time the local bands joined together into tribal camp circles and bison were hunted communally. Tribal relationships were reaffirmed through the performance of rites and temporary social controls were put into effect through men's secret societies called military societies. This was the only time of the year when the tribe was an effective social grouping.

Leadership and authority of the plains tribes varied seasonally with group structure. Bands were led by headmen who achieved their position and who had little authority; they gave advice that was often ignored and depended upon personal influence for leadership. Why? The hunting of scattered bison is an individualistic, not a group, activity and, as we have seen, a loose and fluid organization is well-suited to the uncertainties and vagaries of this type of hunting. Strong leaders are simply not needed for coordination and direction. It was only during raids that strong leadership was desirable and then only for that part of the band that actually made the raid. Thus war parties were led by war leaders who had strong to absolute control. In the summer months, however, strong leadership was necessary to maintain control within the tribal aggregation and, more important, to police the hunt itself to make sure that no individual threatened the success of the group hunt. Authority in the tribal encampments was mostly delegated to the military societies although it was supplemented by tribal officials, either elected formally as with the Cheyenne or through informal consensus as with the Blackfoot.

Status was largely based upon the number of horses that a man owned and his success in warfare rather than upon kinship connections. Good horses were absolutely essential to effective bison hunting. According to Oliver a Blackfoot hunter on a good horse could kill four or five bison on a single hunting trip, while a hunter on a poor horse could kill none at all. Furthermore, horses provided the rapid movement necessary to keep up with the bison. Warfare was another means of achieving status but it was conducted mainly to steal horses from other groups. Although horses were bred by each tribe and some were captured wild, the need for horses was insatiable and was often satisfied by stealing. As in the potlatch of the Northwest Coast, horses were converted into prestige by giving them away to others, thus elevating the status of the giver. For that reason horses were never accumulated in family lines and a young man had to start from the beginning. Horse raiding was a form of competition among the plains hunters. If the horses of a tribe could be kept at a low level, that tribe was at a competitive disadvantage in the never ending quest for bison.

The process of *increasing effectiveness,* the second kind of cultural adaptation, is not easy to study because of uncertainty over what is meant by "effectiveness." One possible meaning is the efficiency with which a food supply is converted into human biomass, that is, thermodynamic efficiency or energy efficiency. Thus, if two human populations *with the same means of subsistence* are compared, the one with the largest size is the most effective. Another meaning of effectiveness is the ratio of food yield to the amount of work expended in its production or exploitation. M. Harris (1971:203ff) refers to such a ratio as "technoenvironmental efficiency" and argues that its value increases with increasing technological complexity. Thus, simple farmers are proposed to produce less food for each calorie of work than advanced farmers. (Whatever the usefulness of this ratio, it must be understood that its designation as "efficiency" is misleading. It is not the same as the concept of thermodynamic efficiency used in ecology. More will be said about efficiency in the following chapter.) Finally, another meaning, and probably the most useful for human ecology, is simply *total food yield,* however efficient its production may be. There can be no question that the "improvements" in food producing technology marking cultural evolution have made larger and larger amounts of food available. Farming with fossil fuels is only the last of these.

Coping with food problems, however, is not the only kind of cultural effectiveness that must be considered in understanding human adaptation. Vayda and McCay (1975) point out that organisms must also cope with *hazards* in the environment. In the case of humans hazards have multiple origins, ranging from geophysical (floods, earthquakes, vulcanism) to biological (disease, epidemics, predation) to social (war, colonialism). How people cope with the occurrence, intensity, scheduling, and duration of such events is a question that must be answered by human ecologists, particularly those working with groups in which the food supply is not limiting. The effectiveness of cultural solutions to those problems, then, must also be measured. Unfortunately, to the best of my knowledge no one has worked on this problem.

The third kind of cultural adaptation, *adaptability,* provides at least some flexibility to cope with unexpected environmental changes. That is, cultural behavior can serve as a "buffer" between humans and random fluctuations in their environment. The social organization of pastoral nomads shows many aspects of a buffering strategy. Louise Sweet's (1965) study of the camel-herding Bedouin of northern Arabia is illustrative. The Bedouin occupy deserts and steppes marked by sparse and unevenly distributed plants and erratic rainfall. Camels, because of their ability to move rapidly over long distances without water, to tolerate heat, and to live on desert plants, are ideally suited for this kind of habitat and have been used by the Bedouin for thousands of years as their primary source of food and transportation. As with the mounted hunters of the North American plains, mobility of social groups is essential in this kind of habitat to take advantage of pastureland when and where it is available. The minimal property-holding and camel-herding unit of the Bedouin is the extended or joint family, and it fulfills this requirement. Such groups normally have around 20 camels, most of whom are females either pregnant or giving milk, and a single tent that can be moved rapidly on baggage camels. The importance of smallness is suggested by the constant friction between individuals leading to the fissioning of these groups as they grow larger. However, small social groups have a critical disadvantage: Their camel herds can be easily wiped out and not so easily built up again by reproduction. The reasons lie in the hazardous physical and social environment of the Bedouin and the exceedingly slow reproductive rate of the camel. Two or three years of drought are sufficient to stop the growth of the all-important desert annual plants and to cause the death by starvation of large numbers of camels. Localized droughts can destroy the camel herds of extended or joint families if migration to better pastureland is not possible. Furthermore, a group's camels are constantly threatened by raids from other groups not only because camels are "prestigious," but also because camels are so difficult to raise. Finally, camels are exchanged at town markets or to itinerant merchants for arms, farming products, textiles, and other supplies. While the threat of losing camels is constant, the possibilities of adding new camels to the herd through reproduction are not good. Camels are bred for the first time when they are 6 years old. After that time, they produce only *one* offspring every 2 years. Even in a large herd only half of these are females and in small herds, such as those kept by minimal families, the proportion of females could be much less.

Sweet argues that all of these disadvantages can be countered by integrating the extended or joint families into a larger social organization, and she suggests that the tribal *section* is the appropriate unit for the Bedouin. The section is the primary residential grouping of the tribe and ranges in size from 20 to 800 tents. Sections are organized into "fixed," localized patrilineages for purposes of camping and into "sliding" patrilineages for purposes of revenge and aggression. The section patrilineages are ranked in relation to each other, and a section chief comes from the top lineage. Communal camel herds are maintained by the section, in addition to the family herds, as insurance against losses; but in the event that a family loses its herds through raids or drought, the section chief assesses other members of the section enough of their camels to replace those lost. The section also takes collective action to protect families from raids, to conduct raids on its own, and to move into adequate pastureland during times of drought, by force if necessary. All in all, the section makes sure that the small family organization can operate effectively to give the flexibility necessary for living in an uncertain habitat.

Whatever the means of adaptation, a beneficial response to an environmental problem cannot be made unless the organism is aware that a problem exists. Many anthropologists see the so-called "humanistic" institutions of human groups playing such a role (e.g., Flannery, 1972b; Rappaport, 1971). Religion, the arts, kingship, codified laws—they all assist in collecting, interpreting, and passing on information so that appropriate responses to the environment can be made. Many of these institutions assign conventionalized "meaning" to the environment and function as "signals" for changes to take place. Roy Rappaport (1968; pp. 237–42) refers to such conventionalization of meaning in the environment as the "cognized" model of the world. The cognized world represents the real world with a set of conventionalized meanings that guide the actions of individuals vis-à-vis the environment. In effect it reduces the amount of information with which the individual must cope. Let us consider the role of *ritual* in information processing. Ritual, according to Rappaport (1971), can best be defined as any *conventional* act of display that conveys information to participants about their current physiological, psychological, or sociological condition. The information conveyed may be part of the *content* or the *occurrence* of the ritual. In what way can the ritual give information about current conditions? David Thomas (1972) provides us with an interesting example from the Western Shoshoni of the North American Great Basin. Although aboriginal Shoshoni rituals were notably scarce, the *fandango* was held regularly and involved a temporary grouping of several local bands, sometimes as many as 300 persons. *Fandangos* varied in content from place to place but nearly always included "gambling, dancing, trading, philandering and courting, and praying for abundance of critical resources" (Thomas, 1972, p. 145). Among the Shoshoni bands that depended upon pine nuts, *fandangos* were held after the pine nut harvest when sufficient food was available to feed a large gathering. The traditional interpretation of the *fandango* is that it was a mechanism of social integration, strengthening kinship ties among the widely dispersed local bands. However, Thomas argues, *fandangos* were also "clearing-houses" of environmental information. Pinyon nut harvests, in particular, fluctuated widely from place to place but could be at least partially predicted a year in advance. "Pooling" the information collected by the local bands would help considerably in predicting *where* next years crops would be abundant and where they would not. Perhaps more important is the *size* of the group attending the *fandango* in relation to the availability of pine nuts, game animals, and so on. Cultural population controls, such as infanticide, senilicide, abortion, and abstinence, could be more stringently applied *if* population size was deemed to be too large for available resources. Otherwise, the "second twin could be allowed to live, the aging grandmother could be nursed along for another year, the illegitimate newborn could be spared" (Thomas, 1972, p. 148).

The *fandango* is not the *only* institution that could function as an information-processor among the Western Shoshoni. In fact it can be argued that the Shoshoni custom of *visiting* is an even better conveyance of information about the state of the environment (Steward, 1938). Members of local bands frequently visited with each other and, presumably, exchanged gossip about local environmental conditions, population size, and so forth. In this case a ritual practice and a social custom, both humanistic institutions, probably join forces to process relevant environmental information.

The occurrence of rituals can be interpreted as a *yes-no* signal, since a ritual either occurs or it does not occur (Rappaport, 1971, p. 26). Therefore, the ritual is a source of qualitative information about the environment if the timing of its occurrence is correlated with an environmental variable. The Courseys suggest that the *New Yam Festival,* common throughout the yam zone in the eastern part of West Africa, can be explained in this way (Coursey and Coursey, 1971). The festival is held each year as the yams begin to mature but must be completed prior to the end of the growing season. Lengthy deliberation by members of the political or religious élite sets the exact date, or the date is set after consultation with the gods. Once the time is established, but before the festival actually takes place, the community undergoes ritual purification, including execution of criminals and making peace with the supernatural. The festival begins with the ritual harvesting of some yams, sometimes from special plots cared for by the priests. These yams are offered to the spirits on the same day that they are harvested, or shortly thereafter, and are then eaten by the community. During the festival feasting, drinking, and sexual activities are indulged in. The Courseys argue that the New Yam Festival is held precisely at the time when the first yams can be cut and still leave enough time so that new tubers can be formed and matured by the end of the growing season. The "early" yams cannot be stored but must be consumed or traded immediately, while the "late" yams can be stored for longer periods of time. However, if the early harvest is too early in the growth cycle of the yam, the plants may be permanently damaged, either producing a very small tuber or dying. Consequently, the lengthy deliberation prior to the festival is necessary to ensure that the timing is exactly right, and the festival itself is a signal to the community to begin the early harvest.

The yes-no signal of the ritual, however, may be used as a simple, unambiguous, means of conveying information about a much more complex quantitative, *more-or-less,* variable or a group of such variables (Rappaport, 1971, pp. 26–27). The *kaiko* ritual of the Tsembaga Maring is a case in point. Extensively studied by Roy Rappaport (1968), the Tsembaga Maring are swidden horticulturalists occupying the northern slopes of the New Guinea central highlands. During his stay from October 1962 until December 1963, Rappaport collected data suggesting that ritual behavior played an important role in regulating a number of demographic and ecological variables, including the number of domesticated pigs kept, the amount of land under cultivation, and the frequency of warfare. Most important of all the rituals was the *kaiko.* The beginning of the *kaiko* signals a period of peace between the Tsembaga and another belligerent, a peace that is ritually celebrated by the planting of *rumbin,* or sacred trees, in each village. When the *rumbin* are uprooted by both parties, the *kaiko* cycle ends and hostilities can begin anew. Uprooting of the *rumbim* requires a festival during which time pig meat must be supplied to all the participants. Rappaport observes that this sets a minimum time limit on the duration of the *kaiko* cycle—that time required for the pig herds to grow to a size sufficient to feed all the participants in the termination festival. However, uprooting of the *rumbim* generally occurs sometime *after* the pig herds are of sufficient size. The exact timing of the festival is a function of increasing information flow from several sources, and the celebration of the festival signals that the sum total of all that information has reached a point where it is possible to make a simple decision to renew hostilities. What are the important sources of information? For one, agitation to uproot the *rumbim* increases with the number of complaints about pigs invading gardens. In turn the number of complaints depends upon the size of the human population and the pig herds in each village. If the human population is becoming large, gardens are spaced closer together and complaints will become more and more frequent even if there are relatively few pigs. On the other hand, a rapidly growing pig herd will cause frequent complaints about garden invasions even if the size of the human population is reasonably small and gardens are widely spaced. Agitation is relieved by celebration of the termination festival. First of all, the large pig meat requirements results in butchering nearly all the pigs in the villages, thus greatly reducing the possibility of complaints about garden invasions. Second, warfare begins soon after the festival, and casualties can reduce the size of the human population. Perhaps a more important mechanism for relaxing population pressure, however, is the capture of enemy lands into which some of the population can migrate.

Physiological Adaptation

A second adaptive level, activated more slowly than behavior, is the reversible and irreversible physiological responses made by the individual. *Acclimatization* is reversible physiological adjustments to environmental stress; none of the changes is genetically transmitted although the potential for physiological change may be. Individual responses to altitude and temperature variation are the best documented examples of acclimatization in humans. If an individual migrates from sea level to 12,000 feet or higher, a typical physiological response is an increased breathing and pulse rate, followed within a few days by an increased hemoglobin concentration in the blood (Lasker, 1969, p. 1481). Temperature acclimatization occurs because humans, like all mammals, must maintain a core body temperature of close to 37 degrees celcius. In cold climates, heat loss is a primary problem and is minimized by reducing blood circulation to the extremities by means of vasoconstriction of the blood vessels and "shivering" to increase food metabolism (however, "cycling" between vasoconstriction and vasodilation is widespread in cold-adapted populations although the phenomenon is unexplained) (Lasker, 1969, p. 1482). Excess heat is the problem in hot climates. Physiological responses include vasodilation of the blood vessels in the extremities to increase heat loss through radiation (the skin surface area of the extremities functions as a radiating surface to the atmosphere) and "sweating" to increase heat loss by means of evaporation from the skin surface.

Plasticity, sometimes referred to as "developmental homeostasis," is an *irreversible* modification in the individual phenotype due to external environmental stress during the process of growth and development. The classic example of plasticity is the large lung capacity and slow skeletal maturation rate documented in individuals living at high altitudes. Both are responses to the low oxygen content of the atmosphere at high elevations. Perhaps a more typical example of plasticity is the increased body size documented in the offspring of Japanese, Swiss, Mexican, and Italian immigrants to the United States (Lasker, 1969, pp. 1484–85). The larger size appears to be a direct response to a new diet. The quantity of food consumed by the offspring is considerably larger than that of the migrants during their youth, and larger body size is an adaptation to a hypercaloric diet.

Genetic and Demographic Adaptation

If an environmental perturbation continues over a long period of time, behavioral and physiological responses are supplemented, if not replaced, by more permanent adaptations. Genetic and demographic changes are the most important. Both have a slow rate of activation, taking several generations, and affect groups rather than individuals. The genetic structure of human groups is maintained with high proportions of favorable genotypes under stable conditions. At the same time a variety of less favorable genotypes is maintained as a "buffer" in case of unexpected environmental change. *Selection* is the process making a genetic shift possible and includes stabilizing or normalizing selection, directional selection, and disruptive selection. *Stabilizing selection* removes very unfavorable genotypes from the group and operates to maintain a status quo. However, *directional selection* changes the genetic characteristics of a group by increasing the proportions of once novel genes at the expense of past favorites. The changes generally take place as a way of improving the environmental fit of the group in the face of changed or changing environmental conditions. Directional selection is illustrated by the relationship between temperature and body shape. Heat loss in humans is largely through the skin (which acts as a radiating surface to atmosphere) and, therefore, the total skin surface area determines the *amount* of heat loss. That is, as surface area increases, so does heat loss. The surface area of any body is directly correlated with its shape: Cylindrical shapes have the greatest surface area and spherical shapes have the smallest surface area for a given volume. In addition, if shape is held constant, the surface area of a body increases in proportion to a reduction in volume. Body shape in humans has a genetic component and, since skin surface area is such an important determinant of heat loss, one would expect selection to increase cylindrical shapes and/or small sizes in hot climates and spherical shapes in cold climates. There is indeed evidence to support the hypothesis. *Diversifying selection* operates upon a group to produce not one but several optimal genotypes. Usually, diversifying selection is an adaptive mode in diverse rather than homogeneous environments. Some parts of the group must solve quite different environmental problems than other parts and, accordingly, undergo different adaptive changes.

Demographic changes cannot all be considered as level three adaptations. The dispersion of a human group over the landscape, for example, can change rapidly in response to an environmental perturbation, such as a volcanic eruption. This kind of modification is really behavioral. However, many demographic characteristics of a group change rather slowly, including birth and death rates, age structure, growth rate, and size. The relationship between the demography of a group and access to essential resources is particularly important in understanding adaptation. R. Brooke Thomas (1975, pp. 72–73) gives the example of a group dependent upon irrigation farming for subsistence. Given a habitat, there is an optimal population size or density that can be supported by the technology. Growth beyond this size would decrease the amount of food available for each individual, causing malnutrition. Similarly, a drop below the optimum size could reduce the labor force to a point at which it could not handle the managerial requirements of irrigation, that is, cleaning silt out of ditches, repairing dams, and so forth. A collapse would follow.

THE ECOLOGICAL SYSTEM

The process of adaptation is dynamic because neither the organism nor its environment remains constant. New problems and new solutions to old problems arise and must be taken into account. What must the organism do to cope with an unexpected earthquake, a raid from its neighbors, a lake that is slowly drying up, or an advancing glacier? In addition, the environment does not have to change to elicit adaptive responses—a previously undetected threat to survival may be suddenly *recognized*. Consider, for example, the recent discoveries of statistical correlations between smoking and lung cancer. Furthermore, organisms change independently of the environment, and those changes may be important to the adaptive process. Genetic mutations or behavioral innovations may present new, and more effective, solutions to old problems. On the other hand, they may cause new problems that must somehow be coped with. All in all, *continual* change in the relationships between an organism and its environment appears to be characteristic of adaptation.

At the same time, it is clear that adaptive relationships are often quite stable and persist over long periods of time. Thus the plants, animals, soils, and so on in a "temperate deciduous forest" maintain distinctive, long-term relationships to each other that persist even in the face of rather drastic shocks. Otherwise, temperate deciduous forests could not be easily separated from other distinctive sets of relationships such as "deserts," "grasslands," or "tropical forests." This stability has led ecologists to define the concept of the *ecological system* (or ecosystem). The ecological system is a set of relatively stable relationships that reflects the adaptation of a group of organisms to each other and to their nonliving environment.

Ecological systems are recognized at different *scales*. On one scale, for example, springs, ponds, cornfields, and the like are microcosms that have all the requisite characteristics of an ecological system. Indeed, most studies of ecological systems have been at this scale. On other scales, the ecological system is identified with large continental deserts, forests, and grasslands, oceans, and even the earth itself. One of the most difficult problems in applying the concept to the study of human ecology is that of scale. Just what is the appropriate scale for understanding the ecological relationships of hunters and gatherers, farmers, city dwellers, and civilized peoples? Is it on the level of an atoll, a mountain valley, a continent, or the earth? The answer depends upon where the group gets its energy and materials and upon how it has coped with hazards, among other things. Furthermore, the scale may change with time, both within a single year or other short period and over a long period of time. Changing subsistence patterns are apt to be responsible for the latter, particularly if the change involves national or international trade.

The different scales at which ecological systems are recognized should make it clear that the concept does not have a reality of its own. Ecological systems are no more than *hueristic* constructs, defined for purposes of understanding ecological relationships, and have arbitrary boundaries. In this sense it is similar to the concept of culture. Nevertheless, some ecologists have argued that ecological systems have distinctive, "emergent" properties in their own right and that they undergo evolutionary changes in much the same way that a species does (e.g., Margalef, 1968).

EVOLUTION

Sometimes ecological relationships are *permanently* changed, a process called *evolution*. There are three distinct kinds of evolutionary processes that may be responsible: *biological, cultural,* and *"emergent"* processes in ecological systems.

Biological Evolution

Biological evolution is permanent *genetic* change, mostly brought about by natural selection. The theory assumes that variation among individuals is in large part caused by genetic differences and is passed on from parents to offspring. Some variants are better able than others to cope with environmental problems and, for that reason, contribute more offspring, on the average, to future generations than do others. The proportion of individuals having the adaptive variant then increases over time. Because of these gradual changes, a group of individuals who normally mate with each other (a Mendelian population) may, if outside mating is prevented, undergo a "genetic revolution." Significant genetic restructuring usually results in reproductive isolation, so that the process is responsible for the origin of new species. The relationship between adaptation and the origin of species underlies an important trend in long term evolution: *diversification*. Diversifying evolution (sometimes called cladogenetic, phyletic, or branching evolution) is the fissioning of old species into new ones that are better adapted. *Adaptive radiation* is the explosive formation of new species following the colonization of a new habitat. The founding ancester is typically an unspecialized species able to cope with a wide variety of environmental problems but not exceptionally well suited to cope with any one. The new species resulting from adaptive radiation are each specialized adaptations to specific problems. But what happens after the new habitat is filled with adapted species? Presumably evolution cannot continue because adaptation is completed; it cannot continue, that is, unless the environment changes or a genetic innovation changes the organism, causing new problems and setting in motion new adaptive solutions. This is usually the way in which adaptation is explained as a continuous process, as mentioned above, and suggests that evolution has no endpoint.

It is almost axiomatic, however, that highly adapted organisms cannot undergo enough change to cope with a greatly changed environment. Rather, like the dinosaurs, they die out. The necessary adaptive solutions to the new problems almost inevitably come from generalized species, able to cope with a wide variety of problems, although not providing very efficient solutions to any of them. For this reason, the course of evolution is zig-zag, shifting from one kind of organism to another, rather than a continuous, unilineal sequence. Since a generalized organism has an evolutionary advantage, it is not surprising to find that the ability to live in a wide variety of environments ("adaptability") has increased over time in organisms dominating evolutionary history. This trend is called *progressive evolution* (or anagenetic evolution).

The theory presented above is often called the "Darwinian" or "neo-Darwinian" theory of evolution because of its stress on adaptation and natural selection as driving forces. Most biologists accept the theory as sufficient to explain the observed diversity and distribution of plants and animals. Nevertheless, others feel that Darwinian theory is not adequate and propose "non-Darwinian" processes to supplement the natural selection of individuals as a way of "sorting out" variants from one generation to the next. This is particularly true of molecular biologists interested in the evolution of proteins. They argue that "most changes in amino acid sequence undergone by several proteins during evolution have been the result of a non-Darwinian process: the random fixation of (selectively) neutral amino acid substitutions" (Arnheim and Taylor, 1969, p. 900). Several human biologists and geneticists have also contended that random fluctuations of selectively neutral genes ("genetic drift") may play an important role in human evolution (e.g., Roberts, 1968). However, studies showing the possibility of random fluctuations as an important evolutionary process assume that genes will be selected *independently* of each other. Proponents of Darwinian theory point out that selection operates on the entire phenotype of the individual, not on separate genes (e.g., Richmond, 1970). At this point it is only possible to say that random changes *may* be responsible for evolutionary change in biological characteristics.

Two other biological processes must also be mentioned as possibly important for understanding the evolution of ecological relationships: *coevolution* and *group selection*. Coevolution was first proposed by Ehrlich and Raven (1964) to explain mutual evolutionary changes in two or more groups of organisms with close ecological relationships, for example predator/prey or host/parasite relationships, but with no close genetic relationships. E. Odum (1971, p. 273) gives the following illustration. Suppose that a plant undergoes a mutation reducing its palatability to a predator, for example, plant-eating insects, perhaps by the production of a noxious chemical compound. Protected from predation, the plant enters a new "adaptive zone" and gives rise to a number of daughter species, all of whom have the mutation. However, the same kind of process can occur in the predators. A chance mutation that enables a plant-eating insect to feed upon the once protected plant group also creates a new adaptive zone for the insect. In the absence of competition with other insects for the plant, the mutated species can then diversify over time. In this case, the evolution of the insect group is dependent upon the evolution of the plant group, and this is what is meant by coevolution.

Group selection, according to E. Odum (1971, p. 274), is "natural selection between groups of organisms not necessarily closely linked by mutualistic associations." Contrary to the Darwinian view of natural selection, differential survival and fertility of local populations and ecological communities, in addition to individuals, are recognized as important evolutionary processes. The essential points of group selection are as follows (Weins, 1966):

1. Groups have some "emergent" characteristics that are missing in individuals.

2. The adaptive requirements of the group often differ from those of the individual; under such circumstances, group needs override individual needs.

3. The adaptive success or failure of entire groups is one kind of natural selection causing evolutionary change.

4. Groups selected in this manner are localized, persist through time, and have ways of reducing gene flow from other groups.

It is not surprising that group selection has been mostly used to explain the evolution of social organization. The selection of individuals alone makes it very difficult to explain the existence of complex social relations in insects, birds, and mammals, especially "altruistic" behavior in which some members of a group willingly die or do not reproduce for the benefit of the group as a whole. Nevertheless, most biologists argue that Darwinian principles are sufficient to explain the phenomena (e.g., G. G. Williams, 1966) or at best accept group selection as a relatively insignificant process. Interestingly enough for our purposes, Alexander (1974) points out that human social groups are almost ideal for the operation of group selection. First of all, local groups tend to be hostile and competitive toward one another. Second, local groups "have been able quickly to develop enormous differences in reproductive and competitive ability because of cultural innovation and its cumulative effects" (Alexander, 1974, p. 376). Finally, "human groups are uniquely able to plan and act as units, to look ahead and purposely carry out actions designed to sustain the group and improve its competitive position" (Alexander, 1974, p. 376).

Cultural Evolution

Julian Steward was the first to demonstrate the usefulness of the concept of adaptation to the study of cultural behavior. Cultural change, in his view, takes place to facilitate adaptation. From this perspective, explaining the evolution of cultural behavior falls logically into Darwinian theory. If adaptation is what causes culture change, then perhaps a process analogous to natural selection is what causes adaptation. Many anthropologists accept this idea and have attempted to bring biological and cultural evolution together under the axioms of Darwinian theory. However, many others object to the analogy, arguing that neither cultural variation nor the "sorting out" process is remotely similar to their biological counterparts. Cultural variations originate in "inventions and discoveries, borrowings, unconscious 'accidents,' political plans, changes from whatever source" (Service, 1971b, p. 10). They are not capable of precise replication and transmission from parents to offspring as are genetic variations; rather they are susceptible to infinite blending and reinterpretation. It is this characteristic of cultural variation that makes a process analogous to natural selection questionable as responsible for evolutionary "sorting."

If it can be assumed that beneficial cultural changes, whatever the source, can be *effectively* taught to offspring then something similar to the selection of genetic variants could occur. The amount of similarity is closely related, or so it would seem, to *how effective* the educational process is. On the other hand, because social and cultural behavior is not confined to the biological boundaries of the individual and can be rapidly transmitted from one person to another, an advantageous innovation is not likely to *gradually* increase among future generations, as it would in genetic selection, but might increase explosively. Governmental control of the educational process, for example, often assures that an innovation will become part of the conventional behavior of all offspring.

Ruyle (1973, pp. 203–204) suggests that an important selection process in cultural evolution is what he calls the "struggle for satisfaction." Individuals have "wants" presumably derived from a combination of genetic inheritance, cultural conditioning, and idiosyncracies. The acceptance or rejection of cultural variants by the individual is determined by how well they satisfy these wants. The fact that advertising psychologists sell products by convincing the public that a product is "desirable" suggests that the process is realistic. In this case, psychological wants are created and are definitely important in the transmission of products and ideas. The transmission of a cultural variant from one generation to the next would, therefore, depend upon *how effectively* the same wants can be created in the offspring, bringing us back to the original point. Cultural evolution undoubtedly involves some kind of "selection" process, indeed probably several kinds, but it is unlikely that something analogous to natural selection is common.

Feedback Control

The processes of evolutionary change in both biological and cultural characteristics are subject to *feedback control*. A change caused by selection, for example, may set in motion processes forcing or encouraging further change ("positive" feedback), resulting in "runaway" evolution. Conversely, the change may start yet other processes bringing about a return to the original condition ("negative" feedback). In adaptation, positive feedback is responsible for continued, evolutionary change, making possible a new set of relationships between an organism and its environment. After a beneficial adjustment has been made, negative feedback provides the "self-regulation" necessary to maintain the relationship.

The way in which negative and positive feedback processes act to control change is illustrated by Flannery (1972b) with data from a small community in Mexico. Until the end of the last century the small, traditional maize farming village of San Juan Guelavia in the Valley of Oaxaca was made up of small property holders with a governing council of elders. Inequalities of wealth were regulated by the ubiquitous Mesoamerican institutions of the *mayordomia* and the *cargo* system. Both institutions involved the exchange of accumulated wealth for prestige within the community. Citizens appointed as *mayordomo* were obligated to assume financial responsibility for the community's patron saint fiestas and other fiestas on the religious calendar. The *cargo* involves the rotation of the governmental offices of the community among the citizens; as progressively higher cargos are undertaken, the prestige and financial obligations of the citizen increases. In theory, citizens who have accumulated the most wealth are appointed *mayordomo* most often and undertake the highest *cargos* so that their wealth is redistributed to the community in exchange for prestige. Thus the *mayordomia* and the *cargo* system are mechanisms for leveling-out wealth differences in traditional Mesoamerican communities.

Toward the end of the nineteenth century Marcial Lopez, a citizen of San Juan Guelavia, was able to convert the *mayordomia* into an institution to increase personal wealth. He was able to do this by "overriding" two critical sources of negative feedback: the church and the council of elders. The church was the guardian of egalitarian values within the community, while the council was responsible for selecting *mayordomos* from those individuals who had accumulated enough wealth to safely assume the financial obligations of the *mayordomia*. Through personal influence, Lopez was able to enlist the aid of the clergy in forcing the council to appoint *mayordomos* from people who were *not* financially capable of assuming that responsibility. Nevertheless, they were obligated to do so because of community prestige and were forced to borrow money. The reader can easily guess from where the loans came: Marcial Lopez! Lopez loaned money with the provision that land be put up for collateral, and, by the end of three decades, he had accumulated much of the community's property through foreclosure. Negative feedback from the church was effectively "shunted" by Lopez' strong support of this institution. As a result by 1915 a few families in San Juan Guelavia, mostly Lopez, owned 92.2 percent of all the arable land and all but 6 percent of the irrigated land. Social stratification had emerged in San Juan Guelavia within a few years, not due to interaction between technology and its environment but due to a positive feedback process affecting ritual institutions.

Mechanisms of Cultural Evolution

Since ecology is the study of of relationships, it is useful to look at several changes in relationships that Flannery (1972b) believes are evolutionary "mechanisms" triggered by the selection and feedback process outlined above. The first mechanism, *the addition of new relationships*, increases the interconnections among organisms and their environment or, from another point of view, among cultural variables. Colonization, empire building, or the creation of tribute alliances such as the Aztec are typical examples. A more recent example is the parasitic foundation of

European agriculture. Imports of fish meal derived from the Peruvian Current off the coast of Peru and Chile are largely responsible for the high agricultural production of Western Europe, particularly in the Netherlands, West Germany, and Belgium (Borgstrom, 1972). Thus in the period from 1966 to 1968, Peru and Chile exported enough fish meal to give 413 million people a minimum protein diet for a year, even though both countries have severe protein shortages. In effect, the output of energy and materials from the Peruvian Current has been coupled to the agricultural system of Western Europe, thus expanding the energy/materials base of that region but reducing the energy/materials base of Peru and Chile.

System relationships can also be *simplified,* as illustrated in the following example (Flannery, 1972b). In the valley of Oaxaca in the Central Highlands of Mexico the control of irrigation canals has traditionally been vested in the autonomous local community. Water is distributed by various officials of the village government who perform the task merely as one of many duties expected of them. Furthermore, the village official responsible for water control does not hold the position permanently; rather, several citizens of the community are rotated into the position by means of the *cargo* system. A final check of the acquisition of too much power in water control is provided by the practice of constructing and maintaining the water network by the community as a whole, along with other community projects such as building roads and schools.

In recent years, however, many communities in the valley have learned of the activities of the Secretaria de Recursos Hidraulicos (SRH), a Mexican federal government agency charged with the development of national water resources. The SRH has specialized personnel and equipment not available to the local villages, and communities in suitable locations have requested this agency to build dams for the storage of water from seasonally-dry streams. Several dams were constructed and local irrigation networks greatly enlarged. However, the acceptance of such a project by a village resulted in the removal of the local village government from the process of decision making in matters related to water. The SRH insisted, because of its financial investment, that its own appointed *agente* be placed in the position of complete control over the local irrigation works. The *agente* is responsible not to the village but to the SRH. Consequently, the once autonomous local communities have been more closely integrated into the larger national political system but have lost the power to make local decisions on water allocation.

Finally, evolutionary change can result by *rearranging relationships*. In the example of the Mexican community of San Juan Guelavia, the role of decision making in the selection of *mayordomos* was switched from the community government and general consensus to the church. (In this example Flannery refers to the change as *promotion,* since it involves the removal of a role from a general-purpose institution, the village government, and its usurpation by a special-purpose institution, the church. However, I am viewing promotion here as a special case of rearrangement since not all such changes will involve a change in the hierarchy of control.) Quite frequently, rearrangement causes the appearance of a *new* part and, if so, it is a mechanism responsible for the evolutionary process of diversification. Thus, in the San Juan Guelavia example, reallocation of roles resulted in the appearance of a "great family"—the Lopez family and their descendants—with unique access to valued resources.

"Emergent" Evolution

Darwinian theory assumes that natural selection, acting upon the individual to improve adaptation, is sufficient to explain evolutionary changes in ecological relationships. However, geneticists and ecologists are inclined to accept other processes as also responsible, several of which have already been discussed. The evolution of "emergent" properties of ecological systems is another process that has been proposed, a process that gives the system "increased control of, or homeostasis with, the physical environment in the sense of achieving maximum protection from its perturbations" (E. Odum, 1971, p. 251). In this case, the ecological system is viewed, for hueristic purposes, as analogous to the organism. Evolutionary changes are then explained by assuming that the ecological system, as well as the individual, must adapt to its environment. As in the individual, the process of adaptation involves changes to minimize the impact of environmental fluctuations. Whether the changes are self-regulating or evolutionary depends, on the one hand, upon how much the environment is changing over time. This conclusion is drawn from the behavior of mathematical models of organisms living in spatially homogeneous habitats that are subjected to random fluctuations over time (e.g., Levins, 1968; R. May, 1974). Under conditions of either unlimited growth or competition, an organism is almost certain to become extinct if the fluctuations are large enough and if the period of time is long enough. A high rate of extinction within an ecological system is really synonomous with evolution. However, an environment varying over time drives organisms to extinction at a low rate compared to *time lag,* that is, delays in the organism's response to an environmental fluctuation (Murdoch and Oaten, 1975, p. 16). This suggests that learned behavior, making possible the most rapid adaptive responses, would tend to be associated with organisms having a low extinction rate. This is certainly true of hominids.

On the other hand, the capacity of an ecological system to remain stable, to resist evolutionary changes, depends upon the *diversity* and *interconnections* of the organisms within.

Diversity

Over the last 20 years, a body of theory has been developed to suggest that *diversity* is the clue to stability, supporting the folkwisdom that too many eggs in the same basket leads to disaster. That is, the greater the number of relationships that organisms enter into, particularly when it comes to their food supply, the less likely it is that an unexpected "accident" will cause severe problems. On the one hand, feeding diversity reduces the chances that a particular organism will suddenly lose its food supply and become extinct; on the other, if an organism does become extinct, plenty of species in an ecological system decrease the probability that the loss will have a significant impact. The evidence thought to support the complexity-stability theory came from observations that (1) simple ecological systems (usually one predator/one prey) in laboratories and modeled mathematically were very unstable and usually became extinct rapidly; (2) agricultural systems with only a few species were subject to explosive pest outbreaks; (3) islands, with rather simple ecological systems, are subject to explosive invasions by foreign species; and (4) the faunas of the arctic tundra and boreal systems, quite impoverished, are subject to violent population fluctuations, while those of the rich tropical forest are not (Goodman, 1975, p. 238). These observations were in turn supported by mathematical arguments, based on information theory, to show that more complex ecological systems would be more likely to survive than simpler ones (e.g., Margalef, 1968).

Within the last 2 or 3 years, however, the validity of the theory has been challenged. In the first place, none of the empirical observations supporting the relationship between feeding or species diversity and stability are beyond reproach. For example Goodman (1975, pp. 238–39) points out the following weaknesses: (1) it has not been shown either experimentally or in mathematical models that complex predator/prey systems with *many* organisms are *more* stable than simple systems; (2) pest outbreaks in agricultural systems have not been allowed to run their course—it is possible that they will eventually stabilize; (3) island "invasions" may be due to chance accumulation of organisms especially susceptible to replacement; and (4) ecological studies in the tropical forests are not sufficiently numerous to show that extreme fluctuations do not occur—indeed there is some evidence to suggest that violent population fluctuations do occur in some tropical organisms. In the second place, the mathematical foundation of the theory is weak. The primary problem is that the Shannon-Weaver statistic, adapted from information theory to measure biological diversity, is next to impossible to relate to field observations and "has no direct biological interpretation" (Goodman, 1975, p. 244). Some other quantitative measure of diversity must be found before mathematical arguments will be convincing.

One kind of diversity, however, has received recent support as an explanation for population stability and, by extension, the stability of ecological systems. Studies of model, experimental, and agricultural systems suggest that stability can be improved by increasing the heterogeneity of an organism's *habitat* (e.g., Levins, 1968; R. May, 1974; Murdoch and Oaten, 1975). Habitat heterogeneity is measured by the number of distinct "patches," that is, resource clusters, available to an organism. Dividing a population into "local" segments, each occupying a separate patch, reduces the probability that the population will become extinct. According to Murdoch and Oaten (1975, p. 10), "since any population owes its existence in part to chance, any form of patchiness, anything that partially uncouples the fate of one segment of the population from that of another, should decrease the probability of simultaneous bad (or good) times for all segments." R. May (1974) has called this "spreading the risk." If a local segment does become extinct, a likely event, migration from other segments can recolonize the patch and establish a new local segment, thereby preventing a drastic fluctuation in the size of the total population.

Although we have focused on a single population, "spreading the risk" has also been observed to increase the stability of predator/prey systems. In this case, patches in the habitat provide on the one hand *refuges* for prey and on the other a mosaic of local predator/prey "subsystems" with more or less independent histories (Murdoch and Oaten, 1975). The stabilizing role of patchiness is helped by anything that keeps patches isolated, particularly barriers to the dispersal of predators. Furthermore, patchiness is a great boon to stability in predator/prey systems *if* the predator feeds on more than one kind of prey and can "switch" from one to another, and if the different kinds of prey occur in different patches (Murdoch and Oaten, 1975). Then, if a prey in one patch is overexploited, the predator can switch to another in another patch, allowing the overexploited prey to recoup its losses.

INTERCONNECTIONS. The contribution of switching behavior to stability suggests an advantage in having numerous pathways between an organism and its environment rather than a few. A vast array of interconnections or couplings opens up *alternative* pathways for energy, materials, and information flow in the event of the blockage of normal routes. In this sense, the organism is "buffered" from unexpected changes in its environment. Up to a point increasing interconnections does stabilize, but there are definite limits. When the couplings become so complex that routes are opened between all or most parts, a change somewhere in the organism's environment is immediately relayed, removing the advantage of buffering. In the case of habitat patchiness, discussed previously, too many connections between patches results in the rapid transferral of local environmental problems throughout the array of local population segments. This "pathological" condition has been called *hypercoherence* (Flannery, 1972b). It is often found in complex political systems where rapid channels of communication are opened between major population centers; the improvement of transportation facilities during the last few hundred years has been particularly responsible for hypercoherence. Consider the example of the Black Plague. After its introduction into the ports of Italy, it spread rapidly into the major population centers of Europe along roads that carried people and freight rapidly from one town to another. The rise of the Black Plague to epidemic proportions was caused directly by hypercoherence, since the removal of some transportation routes would have contained the disease within a few local areas. However, hypercoherence is *not* a general condition; rather, it must be related to a particular problem. Thus while fourteenth century transportation in Europe was sufficiently rapid and complex to cause problems of hypercoherence for epidemic diseases such as the Black Plague, it actually improved stability in another sector of the systems by allowing for rapid movement of foodstuffs from one region to another. A community could now be more effectively buffered from local famine.

SUMMARY

Adaptation is the process of creating beneficial relationships with the environment by means of behavioral, physiological, and genetic/demographic changes. A "beneficial" response is one that contributes to the solution of an environmental problem. Environmental problems may be *evolutionary* in that they must be solved if a genetic or cultural variant is not to become extinct, or they may relate to *wellbeing*, particularly in the case of humans. The latter category includes such problems as general health and happiness, often abstract and difficult to measure but nevertheless important in understanding how humans create environmental relationships.

The process of adaptation in the evolutionary sense involves the "selection" of genetic and cultural variants that are best able to cope with environmental problems and the gradual elimination of those that are not as able. Genetic selection takes place because of differences in the average reproductive success of individual variants, as is well understood. However, differential reproduction is not important in selecting cultural variants, although it may play a role. Several processes of selection are undoubtedly involved, but none are well defined and understood.

Adaptation is a dynamic process because neither the organism nor its environment are constant. New problems are continually arising and new relationships established to provide solutions. Nevertheless, adaptive changes usually take place within limits, maintaining distinct organism/environment relationships over time. The *ecological system* is the result of adaptive stability, reflecting no more than a distinct set of relationships that persists over time among a group of organisms and their nonliving environment. Just how stable an ecological system is depends upon the interconnections among its organisms, the heterogeneity of the habitat that they occupy, and the domination of negative feedback processes. Instability, on the other hand, follows from the domination of positive feedback processes, habitats that fluctuate randomly over time, and organisms that make slow responses to environmental changes. The evolutionary transformation of adaptive relationships arises out of instability.

THREE
ENERGETICS

Of all the problems of adaptation, energy has received the most attention from ecologists. Many years ago the late Raymond L. Lindeman (1942) demonstrated that the study of how organisms cope with energy problems can go a long way toward explaining ecological relationships. He proposed an "energetics" model in which organisms were grouped together into "trophic" or feeding levels and then reduced to their heat energy (calories) equivalent. Feeding relationships among trophic levels could then be measured by the number of calories exchanged. Not only did the calorie provide a *common denominator* for quantifying interactions among organisms but also it made the interactions explainable by physical laws, the laws of *thermodynamics*.

PRINCIPLES OF THERMODYNAMICS

The basic behavior of energy in an ecological system is governed by two laws of *thermodynamics*. First is the *law of energy conservation*, stating that energy is neither created nor destroyed but may be changed from one form to another. Consequently, every ecological system has a balanced energy budget—what goes into the system exactly equals what comes out. Nevertheless, numerous energy exchanges and transformations may take place so that balancing an energy budget can be a complicated task indeed. The second law of thermodynamics is the *law of entropy* and states that in a spontaneous process (which does not require an energy input from the environment) energy is continually degraded from a more organized to a less organized form. Since heat (random molecular movement) is the most unorganized form of energy, the law of entropy also means that all forms of energy are eventually converted to heat and lost to the environment. The law is manifested by heat loss each time an energy exchange or transformation takes place; for example, a predator eats a prey or solar energy is changed to chemical energy by photosynthesis.

THE FORMS OF ENERGY

The first law of thermodynamics suggests that energy is found in many different forms, and this is indeed the case. An accurate assessment of energy in any system must take all of these into account. The principal forms of energy are chemical, mechanical, heat, electromagnetic, nuclear, and sound. (The following discussion is based upon Miller, 1975; Moran, Morgan, and Wiersma, 1973; and Phillipson, 1966.)

Chemical energy is stored in the molecular and atomic structure of many complex substances. The breakdown of complex molecules due to respiration processes is typical, and the oxidation of glucose to carbon dioxide and water is typical of this kind of energy:

$$C_6H_{12}O_6\text{(glucose)} + 6O_2 \rightarrow 6H_2O + 6CO_2 + \text{heat}$$

The burning of fossil fuels such as coal, another oxidation process, is an important source of energy in modern industrial societies.

Released chemical energy is often converted to *mechanical* energy. Mechanical energy, unlike the other forms of energy, can do *work,* which is defined simply as the product of *force* times the *distance* over which the force is applied. There are two kinds of mechanical energy: potential and kinetic. *Potential* energy is energy at rest or stored energy, and becomes *kinetic* energy as soon as it begins to move. The energy of motion, or kinetic energy, is closely related to *heat* or thermal energy, since by definition heat is a measure of the average energy of the motion of molecules. Heat energy is commonly measured in terms of the *calorie* (or gram calorie) (the amount of heat required to increase the temperature of one gram of water from 3.5 to 4.5 degrees celsius), or the *large* (or kilogram) *calorie* (1000 calories).

Electromagnetic energy is in the form of *waves* whose length and frequency vary. The shortest waves and the waves with the highest frequency have the highest energy content. Red and blue light emanating from the sun is perhaps the most important form of electromagnetic energy for our purposes, since it is essential to the process of *photosynthesis*. Photosynthesis takes place in green plants (which contain the green pigment chlorophyll) and converts substances with low chemical energy to substances with high chemical energy. The process is as follows:

$$6CO_2 + 6H_2O + \text{light energy} \rightarrow C_6H_{12}O_6 \text{ (glucose)} + 6O_2$$

The stored chemical energy in glucose can then be converted into kinetic mechanical energy by the process of respiration. For most purposes the glucose manufactured and stored in green plants is the energy foundation upon which all life on earth depends.

Several kinds of electromagnetic and heat energy are implied by the term *nuclear energy*. Nuclear energy is produced by either splitting the nucleus of atoms by bombardment with neutrons (nuclear fission) or by fusing together small atomic nuclei into larger ones by the application of very high temperatures (nuclear fusion). Both processes result in the emission of high-energy gamma rays and other forms of electromagnetic waves, along with heat.

Finally, *sound energy* must be mentioned, although it is not an important source of work and power. Sound energy is similar in many ways to electromagnetic energy, since it occurs in the form of waves; however, unlike electromagnetic waves, it cannot be radiated in a vacuum and must have a medium such as water or air. Sound waves have very low energy content. Nevertheless, the impact of sound waves upon the quite sensitive human ear can be dramatic, to which anyone who has gone to a rock concert can attest.

THE MECHANISMS OF ENERGY EXCHANGE

How organisms exchange energy, in whatever form, is the fundamental question of ecological energetics. Processes or "mechanisms" of exchange are the same as for any object and conform to physical laws.

Physical Mechanisms

Organisms exchange electromagnetic and heat energy with the *nonliving* environment by means of radiation, conduction, convection, and evaporation (Gallucci 1973).

Radiation is the direct movement of energy "waves" from and to an object. The rate of radiation depends upon the properties of the object's surface (e.g., composition or color), the surface area, and the kind of energy being radiated (e.g., wavelength).

Conduction is energy exchange between objects or an object and a fluid (e.g., air or water) in direct contact (and at different temperatures). The rate of conduction depends upon the "insulating" properties of the objects or the "conductivity" of the fluid.

Convection is energy exchange between an object and fluids (e.g., air, water) moving over its surface. The rate of convection depends upon the fluid's speed and the difference in temperature between the fluid and the object.

Evaporation is energy exchange caused by the conversion of a liquid to a gas or vice versa. Organisms lose energy by transpiration in plants and evaporation or sweating in animals (each gram of evaporated sweat removes 0.58 kilocalorie of heat from the animal's surface). An organism gains energy by the condensation of moisture from its fluid surroundings.

Each of these exchange mechanisms makes it possible for the organism (or any object) to have a dynamic relationship with its environment. Temperature differences are especially important. If the organism is hotter than its surroundings, radiant and heat energy are lost; if colder, energy is gained. For "homeotherms", organisms that maintain relatively constant body temperatures, energy losses and gains through radiation, conduction, convection, and evaporation are critical. A variety of physiological, genetic, and behavioral responses to environmental conditions regulate the amount of energy exchange that takes place through each mechanism. Increased sweating when body temperature gets hotter, for example, is a physiological response that increases heat loss through evaporation. Several "ecogeographical" rules reflect genetic responses to the same problems. For example Bergmann's rule (body shape is a function of climatic temperature) and Allen's rule (the length of body limbs is a function of climatic temperature) both state that body surface area varies directly with the temperature of an organism's surroundings. Since larger body surface areas increases the rate of radiation to the environment, thereby cooling the organism, its association with hot climates is not surprising. The reverse would be true for cold climates, in both cases suggesting an adaptive relationship.

Feeding Mechanisms

Organisms exchange biomass (chemical energy) with their *living* environment by means of *feeding* (Gallucci, 1973). Feeding as an exchange mechanism is manifested in *food chains* or *food webs*. The simplest is the food chain in which each "link" in the chain is a single feeding relationship between two organisms. Feeding ultimately starts with organisms that are capable of photosynthesising solar radiation, usually green plants, and are called *autotrophs* ("self-feeders"). The next link is *herbivores* ("plant feeders"), made up of organisms that feed directly upon plants and therefore need only a single exchange of biomass. First-level carnivores ("meat feeders") feeding upon herbivores, are next in a food chain, followed by second-level carnivores, third-level carnivores, and so forth).

Ecologists actually define two kinds of food chains: the *grazing* food chain and the *detritus* food chain. The grazing food chain starts with green plants and is common to all organisms who feed upon *living* matter. Not all organisms feed upon living matter, however. The detritus food chain begins with the waste and dead matter sluffed-off at each link in the grazing chain. *Saprotrophic* ("decomposition eaters") organisms, such as bacteria, yeasts, and molds, decompose dead organic material into inorganic substances and are thus critical to material cycles in the ecosystem. *Detritovores* (e.g., mites, snails, beetles) subsist upon dead organic matter that has been "sluffed off" (e.g., leaves) and upon saprotrophs. Finally, a series of higher-level carnivores feed upon the detritovores. Energy flow in some ecosystems is almost exclusively through

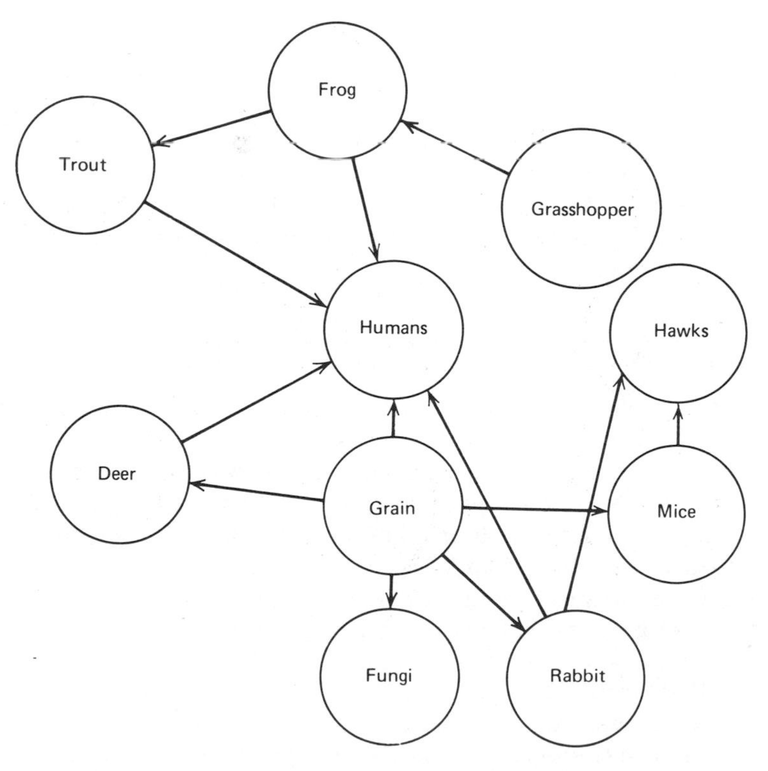

FIGURE 3.1 A simplified food web in which humans participate.

detritus food chains, for example, springs in a deciduous forest, and the role of detritus chains is important in most ecological systems.

Feeding relationships are seldom simple. Rather, several chains are interconnected to form a *food web*. Any population is usually immersed in notoriously complex food webs that often change seasonally. Nevertheless, the links making up the web can often be approximated, as illustrated in Figure 3.1. A food web can be described by the number of its "links" and "trophic levels." A *link* is one energy transfer, or to put it another way, a connection between an organism and a single source of food. The larger the number of links in a food web, the more complex the food habits of the organisms linked by the web. *Omnivore* is a term commonly applied to organisms with a wide range of food habits, those that eat both plant and animal foods, and omnivores are likely to be part of food webs with many links. Man is the omnivore with which you are most familiar. A *trophic level* groups together organisms who feed on food chains approximately the same distance away, or the same number of energy transfers, from the beginning—either solar radiation or debris. The common trophic levels are producers (autotrophs), primary consumers (herbivores), secondary consumers (first level carnivores), etc. Therefore, a food web that includes producers and primary consumers only has two levels; if it also includes secondary and tertiary consumers, it has four levels. It is important not to confuse a particular *species* of organism with a trophic level, since the same species can occupy several trophic levels. For example, a human population feeding upon wild seeds, deer, and carnivorous fish is at once a primary consumer, secondary consumer, and a tertiary consumer.

THE EFFICIENCY OF ENERGY EXCHANGE

The law of energy conservation requires that each time energy is exchanged what goes in must exactly equal what goes out. That is, the energy budget of any object or system must be balanced. The second law of thermodynamics, the law of entropy, introduces a complication into the balanced equation. Each time energy is exchanged some degradation to *heat* occurs. In food chains feeding exchanges are controlled by the same law. Organisms cannot make use of heat energy as a source of food so that less food is available in a food chain each time an exchange takes place; ultimately, the chain is brought to an abrupt halt. For this reason energy flow through food chains, trophic levels, or ecological systems is said to be *one way* rather than cyclical. The one way flow is effectively illustrated by the biomass (chemical energy) content of each step in a food chain. Plants have a relatively large content, herbivores only about 10

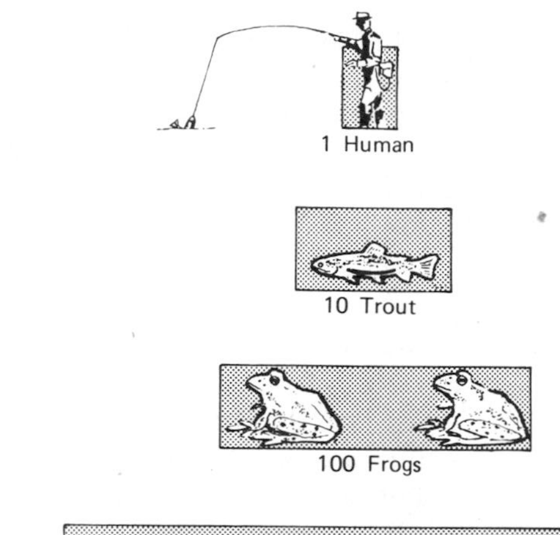

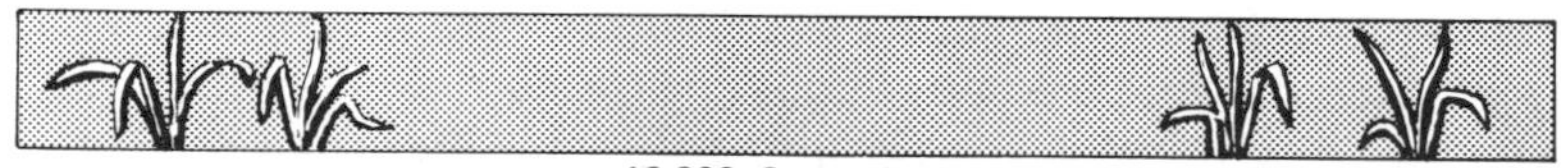

FIGURE 3.2 A population pyramid.

percent of that, first level carnivores about 10 of what herbivores have, and so forth. Figure 3.2 shows the relationship in a hypothetical chain. Without a continuous flow of energy into the chain from an outside source (ultimately the sun), it would rapidly collapse.

The energy lost during food chain transfers is measured by an index of *thermodynamic efficiency* (Kozlowsky, 1968). In its most general sense efficiency is the ratio of the energy content of an organism to that of its food supply. For example the efficiency of an herbivore is calculated by dividing its biomass (converted to calories) by the biomass (also converted to calories) of the plants that it eats. The ratio is then changed to a percentage (multiply by 100). In food chains there are two important sources of energy loss. The first is feeding, the actual transfer of energy from one individual, population, or trophic level to another. Energy losses take place during feeding because not all available food is found, captured, and consumed and because not all consumed food is assimilated (absorbed or metabolized). *Consumption* losses are due to the ineffectiveness of feeding "strategies," failure to find or capture food that is available. In human groups consumption losses are also due to food preparation methods (e.g., butchering and cooking practices) and to food values (e.g., Western dislike of animal blood, an eatable food that is used by many human groups). *Assimilation* losses have metabolic origins but also depend upon the type of food being eaten. According to Engelmann (1968), poikilotherms (organisms that cannot regulate body temperature) assimilate only about 30 percent of what they eat, while homeotherms assimilate about 70 percent. Of the homeotherms, planteaters lose more to waste than do meateaters.

The second source of energy loss is *within* the same trophic level. Respiration, the chemical breakdown of assimilated food into usable energy, is a metabolic process necessary for life support activities, including feeding, reproduction, cooling and warming, and so forth. Large energy losses are caused by respiration, as predicted by the law of entropy. For example 10 to 20 percent of the glucose photosynthesized by plants is lost to heat. In animals heat loss from respiration is even greater: invertebrates lose 79 percent and vertebrates lose 98 percent! With losses that heavy it is easy to see why food chains are always short. Animal eaters have slim pickings indeed. Only the energy left *after* respiration can be used for *production*, stored in biomass by growth or reproduction. Only production can be eaten, that is, passed on to organisms at higher trophic levels or to other food chains (e.g., detritus chains).

PRODUCTION

The energy flowing into food chains and ecological systems ultimately comes from solar radiation and is dependent upon a single exchange—solar radiation/plants (or other autotrophs). Only autotrophs can photosynthesize the biomass used in all other food chain transfers. *Primary productivity* is the rate at which autotrophs produce biomass and is, therefore, a measure of the energy available to a food chain or ecological system. Table 3.1 shows plant productivities before respiration (*gross* primary productivity) in a range of ecological systems. The percentage of solar radiation that is used during production ("efficiency") is also given. Comparison of the figures in the table suggest some important similarities between "natural" and "artificial" (e.g., farming) ecological systems. Photosynthetic efficiency is low in both types, placing severe limitations on energy flow through the system and through its constituent food chains. Human technology has not improved the situation. Although productivity rates are more variable than efficiencies, "artificial" systems do not have rates that are high compared to many "natural" systems. Again, while human technology has been able to circumvent some limitations on gross production (e.g., water, nutrients), it has not been able to significantly improve upon nature. All food chains and ecological sys-

TABLE 3.1 GROSS PRIMARY PRODUCTION IN SEVERAL NATURAL AND CREATED ECOLOGICAL SYSTEMS[a]

Ecosystem	Production (kcal/m²/day)	Efficiency (percent sunlight used)
I. Systems with Little Production		
Deserts	0.4	0.05
Arctic tundra	1.8	0.08
Subtropical blue water	2.9	0.09
II. Fertilized Artificial Systems		
Tropical forest plantation	28	0.7
Algal culture	72	3.0
Sugar cane	74	1.8
III. High-Yield Natural Systems		
Coral reefs	39–151	2.4
Tropical rain forest	131	3.5

[a] *Data from H. Odum, 1971, p. 83.*

tems, whether created by humans or not, are "driven" by very similar *rates* of energy flow.

Of course, consumers, such as humans, cannot eat gross primary production. Only that remaining after respiration, *net* primary production, can be used by the rest of the food chain, and it is here that human technology has had the most success in improving man's food supply. In "natural" systems with very high rates of gross primary production, respiration losses are also extremely large. For example a mature tropical rain forest in Puerto Rico has been shown to lose over 71 percent of its photosynthesized energy to respiration (E. Odum, 1971, Table 3-5). By contrast "artifical" farming systems lose much smaller amounts. Thus, a highly productive alfalfa field in the United States lost less than 38 percent to respiration, leaving the rest to be harvested. Eugene Odum (1971, p. 48) summarizes the important differences between "natural" and "artificial" systems by stating that "nature maximizes for gross production, whereas man maximizes for net production." Perhaps the most common method used by humans to achieve that goal is to prevent ecological systems from reaching "maturity" because "immature" systems have relatively low respiration losses. (Chapter 5 will discuss the methods in detail.)

MODELS OF ENERGY EXCHANGE

Although energy exchanges within an ecological system are enormously complex, they can be reasonably represented by *models.* Models are simplified portrayals of the "real" world using pictorial, verbal, or mathematical symbols. They differ from each other in degree of *realism* (how accurately the model depicts real objects and processes), *precision* (how accurately the model predicts the behavior of the real world), and *generality* (the number of real world situations to which the model can be applied) (Walters 1971). The purpose of the model determines in what way the real world is represented.

Models of energy exchange are often constructed with *compartments* to represent organisms, trophic levels, solar radiation, respiration, inorganic materials, and so forth. The *energy content* of each compartment is then specified by a constant or a mathematical function. Energy exchanges between compartments are represented by connecting *pathways* showing the direction and *rate* of energy flow. Transfer rates are given either as constants or as mathematical functions. Figure 3.3 illustrates a very simple energy exchange model.

It is important to remember that an energetics model *must* have a balanced energy budget; that is, the energy flowing in must equal the energy flowing out, in whatever form it may be. The law of energy conservation demands that this be true. In human groups input energy is likely to be both solar radiation and imports (e.g., debris, traded food, or fossil fuels). Subsequent energy exchanges *via* feeding must have losses to heat (output energy) and to waste (output energy to other food chains or, if used for fuel, converted to heat energy). Within the human group, life support activities, particularly the work involved in feeding, has large heat losses from respiration (output energy). What energy remains (production) is in the form of biomass (chemical energy) and is on the output side of the budget. Energy exported outside must also be tabulated, as must all energy exchanges *via* the physical mechanisms of radiation, conduction, convection, and evaporation.

Although an energy budget must be balanced over a relatively long period of time, a temporary *imbalance* can occur over short periods. In humans energy flow is often significantly different from one season of the year to the other, one season showing more energy loss than gain, the next showing the reverse. The imbalance is generally reflected in ups and downs of individual body weight, as supported by a study of farmers in Gambia, West Africa (Harrison et al., 1964, pp. 428–33). Food during a normal year is at a maximum from the end of November, when harvesting takes place, until the end of February. Although the harvest is stored, food consumption is at a peak during this period and labor demand is at a yearly low. The excess of intake over energy expenditure is reflected in the rapid gain in individual weight. Supplies are decreasing from March until May so that food consumption falls off; however, there is still a low labor demand. Body weight tends to be constant during this time because

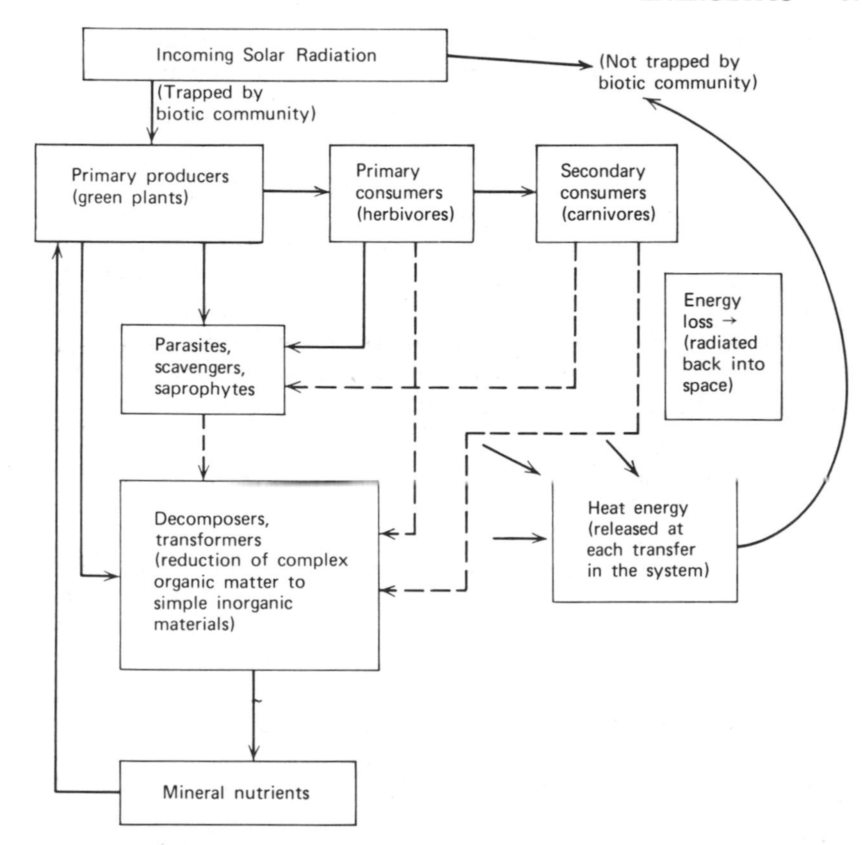

FIGURE 3.3 Compartmental model of the energetics of an ecological system. (From *Human Environments and Natural Systems: A Conflict of Dominion* by Ned H. Greenwood and J. M. B. Edwards. Copyright 1973 by Wadsworth Publishing Company, Inc., Belmont, Calif. Reprinted by permission of the publisher, Duxbury Press.)

of a balance between low intake and low expenditure. Food supplies and consumption remain low during the period from May until September but energy expenditure rapidly increases and reaches an annual peak because of the advent of a new planting season. The inequity between intake and expenditure results in a rapid fall in body weight. Finally from September through November weight loss is less rapid because of a decrease in labor. However, food consumption is at a yearly low so that weight continues to drop. At the end of November the harvest is in and the cycle begins anew. Seasonal fluctuations in energy flow, then, force the field worker to pay attention to problems of *sampling* over time.

Energetics data must be collected during a representative number of time periods, say winter, spring, summer, and fall, not just from one period.

HUMAN ENERGETICS

From one point of view, the study of human energetics is no different than that of any other omnivorous creature. The mechanisms of energy exchange are exactly the same, and thermodynamic efficiencies are not greatly different. What amount of energy actually flows to the human group depends upon its position in a food chain, as it does with any other animal. However, from another point of view, the study of human energetics is unique for the simple reason that the fundamental characteristic of human adaptation within the last 10,000 years is the *manipulation* and *creation* of the feeding relationships. The complexity of the change is enormous. Food chains have been greatly simplified and shortened by removing organisms that compete for the limited input energy. Gross primary productivity, the input energy, has been improved in some cases by removing severe physical limitations, such as water or nutrient extremes. The percentage of input energy available for human consumption, net primary productivity, has been kept at a maximum by maintaining "immature" ecological systems. The usefulness of food species has been improved by genetic manipulation so that more of their energy goes into parts that humans find desirable. (Chapter 5 discusses these changes in more detail.)

Efficiency of Feeding Methods

Human restructuring of food chains suggests that the only way to evaluate the effectiveness of their feeding methods is to measure the proportion of energy entering the food chain that is actually transferred to the human population. One such measurement is:

$$\frac{\text{food yield}}{\text{solar energy} + \text{imported energy}}$$

Energy flowing into the chain from the environment comes from solar radiation striking the earth's surface either *within* a group's geographical boundaries or *outside* but imported to the group. Traded foodstuffs and fossil fuels (representing photosynthesized solar energy from an ancient food chain) are typical examples of imported energy. The *total* energy coming from these sources, expressed in calories, is the denominator of the equation. If solar energy values are given as a *rate* per unit area (e.g., calories/ square meter/year), they must be multiplied by the total area of the group's habitat to give a "habitat rate." For example, if the rate is 100 calories/ square meter/ year and the habitat size is 10,000 square meters, the habitat rate is $100 \times 10{,}000 = 1{,}000{,}000$ calories/year. The rate of imported energy (e.g., imports/year) is added to the habitat rate to give the total input. Only a portion of this, however, is transferred to the human population, the remainder being lost to heat, other organisms, ineffective feeding strategies, exports, ritual destruction, food preferences, and a variety of other factors.

It is important to remember that the measurement of the food *yield* of a cultivated crop or a hunted animal does not necessarily indicate energy transferred to humans, particularly if a large part of the crop is ritually destroyed or exported. *Consumption* is the only real measurement of energy transferred to a human group, after it has been corrected for assimilation losses (body waste). Accordingly, a second measure of feeding effectiveness is:

$$\frac{\text{assimilated food}}{\text{solar energy} + \text{imported energy}}$$

Collecting data on food consumption is a time-consuming task and one that is plagued with sampling problems. It is obviously impractical to collect information from everyone in the group so that a *representative* sample must be taken, including all age groups and sexes. Furthermore, the time period during which data are collected must be sufficiently long to correct for seasonal and/or yearly variations. Obtaining permission to do consumption studies is often difficult. Questions about the number of yams eaten last month are likely to meet with the same responses in New Guinea as an inquiry about take-home pay in a small West Virginia town. Rappaport (1968, Appendix 9) observes that it took him two months to gain the cooperation of the Tsembaga Maring (New Guinea) in conducting studies of this type, and an equivalent, or longer, period of time could be expected almost anywhere.

Both of the above equations are, of course, measures of feeding "efficiency." One way to solve the problem of food acquisition, is to increase feeding efficiency but, interestingly enough, human history is not characterized by revolutionary advances in feeding efficiency. Rather, feeding improvements have come about by increasing the *volume* of energy flowing into food chains, even at the expense of reduced efficiency. One common way to increase volume is to expand the geographical *scale* of a group's habitat, giving a larger solar energy base.

Technological modification of the earth's surface, for example, irrigation, swamp reclamation, or fertilization, allows expansion into previously unoccupied regions but *political domination* of occupied regions, including conquest and trade, is an equally common means of enlarging the energy base of a population. The use of *fossil fuels*, an imported source of energy, is another way in which the volume of the energy base has been increased. Modern farming consumes huge quantities of fossil fuels, making it the most inefficient feeding method ever devised by man. Nevertheless, enough of the fuel is, in effect, converted to food that modern farming also has the largest yields. The modern world eats "potatoes partly made of oil" (H. Odum, 1971, p. 115). The dilemma, of course, is that the world reserves of fossil fuels are being rapidly depleted, eventually forcing a return to complete dependence upon solar energy or to newly developed sources such as nuclear energy.

Cost-Yield Ratios

The fate of energy after assimilation is another part of human energetics that must be brought to bear on the effectiveness of feeding methods. The reader should recall that *life support* and *production* are the two destinations of assimilated energy. By far the most work has been done on the life support requirements of human feeding methods. For example, in recent years several anthropologists have collected data on the relationship between food yield and the work done to produce that yield (e.g., M. Harris, 1975; Lee, 1968; and Rappaport, 1968). The maintenance "cost" of a feeding method is measured by a *cost-yield ratio*, the calories of food yield for each calorie of work expended in its production or exploitation. Foragers have been calculated to have a ratio of 9.6:1 (!Kung Bushmen, Africa), hoe farmers a ratio of 11.2: 1 (Genieri, Africa), slash-and-burn farmers a ratio of 18:1 (Tsembaga Maring, New Guinea), and irrigation farmers a ratio of 53.5:1 (Luts'un, China) (M. Harris, 1975, pp. 234–245). Harris concludes from this that the *relative* cost of food has decreased during the course of human evolution.

ABSOLUTE COST. Although the relative cost of food production has decreased, the *absolute* cost has increased (e.g., M. Harris, 1975). That is, feeding methods that produce larger yields require more work for their maintenance. Boserup (1965), for example, calculates that slash-and-burn farmers have low food yields but that each family works only about 1000 hours a year. By contrast intensive farmers raising several crops a year have high food yields but each family works 5000 hours a year. M. Harris (1975) concludes that the evolution of human feeding methods has made it possible to work more, not less. Nevertheless, Boserup (1965) argues that, given a choice, a human group will always pick the feeding method that has the lowest absolute cost rather than the one with the highest yield. Although this position is controversial (cf., Bronson, 1972), is suggests that the proper evaluation of human feeding methods must consider the possibility that *low cost* is as advantageous as *high yield* under some circumstances.

What these circumstances are is outlined by Schoener (1971) in a general discussion of feeding strategies in ecology. According to this author, although the maximization of food yield would seem to be adaptive, an attempt to do so introduces other adaptive problems that may make it disadvantageous. The main problem is the limited *time* available to an organism for all of its life support activities. An attempt to maximize food yield will require larger and larger amounts of time expended in food-getting activities. Other critical activities, such as the search for mates and mating behavior, must also take place within that limited time, and increasing feeding time (cost) necessarily impinges upon those activities. Therefore, how much time is allocated to each activity varies with its importance. Among social animals that depend upon learning, for example, time spent in play, in watching adults, in correcting the young, or in the classroom is as important as the time spent in feeding. Feeding time will suffer accordingly. Furthermore, Schoener observes that if an organism is subject to predation, feeding usually means coming out of hiding places and being exposed to the risk of being eaten. Cutting down on feeding time may be necessary as protection against heavy predation losses; indeed, *minimizing* feeding time may be advantageous. Feeding strategies, then, are likely to be *optimal* solutions to competing time problems and are subject to selection. Schoener suggests that selection has produced two fundamental feeding strategies: *time minimizers* and *energy maximizers*. Time minimizers reduce feeding time to the absolute minimum to favor other activities, while energy maximizers increase feeding time to the absolute maximum at the expense of other activities. Returning to human feeding strategies, Boserup is suggesting that humans are time minimizers. Some support for her position can be found in the evidence for heavy predation on early hominids, along with hunting as an important means of feeding. Hunting is an activity that involves movement away from the group or hiding places for relatively long periods of time. Predation is bound to have taken a heavy toll of hunters, perhaps selecting for a feeding strategy that minimizes the amount of time spent in hunting. On the other hand, feeding strategies of this type may be more closely tied to population growth characteristics. For example, Rapport and Turner (1975) argue that energy maximizers are associated with rapidly growing populations, while time minimizers are associated with stabilized populations. This interpretation fits Boserup's data just as well, since increasing food yield and cost are explained by "population pressure." Population pressure in her context means no more than a rapidly growing population.

TIME-MOTION STUDIES. Precise data on the maintenance costs of human feeding methods can be obtained with modern physiological equipment (e.g., R. Thomas, 1973). However, when it is impossible to use specialized equipment, an *estimate* of respiration losses can be made by conducting *time-motion* studies. The method requires keeping a time record of the different activities ("motions") engaged in by a human group. For example, a stopwatch is used to measure the time spent in each of several activities by adult men, adult women, and children (perhaps also broken down by sex if different activities are involved). Such a time-motion study of one activity is given by Rappaport (1968, Appendix 9) in his work on the Tsembaga Maring of New Guinea. The study was conducted to measure respiration loss during the clearing of a one-acre tract of forest.

First, the number of *movements of the same type* are recorded for a selected time period. For example, if an individual is cutting a tree with an axe, each stroke of the axe is assumed to require approximately the same amount of energy expenditure. Counting the number of strokes used to cut the tree thus provides a measure of total energy expended for this activity. If the trees are about the same size, simple tabulation of the trees cut multipled by the number of strokes required to cut a single tree will give an estimate of energy expenditure for clearing trees from a field. A more accurate estimate can, of course, be made by counting the total number of axe strokes used to cut *all* the trees. This procedure corrects for variation in tree size but requires a lot of time on the part of the anthropologist.

Since *several* individuals are normally involved in the same activity, the size, sex, and age of the individuals must be recorded. Each of these factors may affect the amount of energy expended per movement. (See Table 3.2.)

To complete the time and motion study, the field worker must have access to data giving the number of calories expended (respiration) per unit of time for each type of movement and for each body size, age group, and sex. Such data are available for individuals in modern Western civilization but are limited for other societies. (See Table 3.3.) Normally,

TABLE 3.2 CLEARING UNDERBRUSH: A TIME AND MOTION STUDY

Worker's Name	Sex	Estimated Age	Weight in Pounds	Time	Number of Strokes	Time	Number of Strokes	Time	Number of Strokes	Comments
Akis	M	20	88	10:37–10:43	296	11:14–11:20	250	12:14–12:20	248	Only one 3-min break during period. Next longest break: 15 sec.
Acimp	F	50	85	10:55–11:01	244	11:30–11:35	209			No breaks longer than 20 sec during working period.
Avoi	M	55	94	11:02–11:08	177	11:50–11:56	190			Slower than other workers because of short breaks and longer strokes.
Men	M	28	120	6 min	233					Longer strokes than any of the others.
Wale	F	35	76	9:53–9:59	246	10:53–11:04	260			No breaks longer than 20 sec during working period.
Nimini	M	18	96	6 min	246					
Mer	M	40–45	94	6 min	316					Stated that he was in a hurry.

Source. Reprinted by permission of Yale University Press from *Pigs for the Ancestors* by Roy A. Rappaport. Copyright © 1968 by Yale University.)

TABLE 3.3 ENERGY EXPENDITURE FOR DIFFERENT ACTIVITIES. THE SUBJECT WAS A YOUNG MALE OFFICE WORKER IN BRITAIN

Activity	Time (min)	Energy Expenditure (kcal/min)	Total (kcal)
Sleeping and lying in bed	457	1.3	594
Sitting	620	1.6	992
Standing	125	2.25	281
Washing and dressing	42	2.9	122
Domestic work	70	3.0	210
Walking	96	5.6	536
Cycling	30	6.4	192
	1440		2927

Source. *The Biology of Work* by O.G. Edholm. Copyright © 1967 by O. G. Edholm. Used with permission of McGraw-Hill Book Company.

even if these data are available, differences in physical characteristics between the control group and the group being studied will introduce some error. In Rappaport's study respiration figures were taken from data already published on other New Guinea groups and applied to the Tsembaga, a procedure that is questionable but often unavoidable.

Production Efficiency

What assimilated energy is not expended in maintenance is used for *growth* and *reproduction*. It is the net production available for transfers to other individuals, populations, or trophic levels. For example, wild cats and dogs preying upon early hominids had this proportion available to them, as do today's "man-eating" tigers, crocodiles, and sharks. In a less exciting vein, net production is also the proportion of energy available to detritus chains after a human death. A final measurement of the efficiency of human feeding methods, then, is the proportion of energy entering the food chain that is actually converted to *human biomass*. That is,

$$\frac{\text{human biomass}}{\text{solar energy} + \text{imported energy}}$$

DOES MAN LIVE BY CALORIES ALONE?

Energy flow models based upon calories do not tell the whole story, although the ecological literature can be searched far and wide before finding any other kind. The problem is that an organism may have plenty of calories but still not survive because it lacks other essential things from its environment; indeed, in today's world most human starvation is caused by not enough protein, not by a lack of calories (Berg, 1973). In man the individual must have *carbohydrates* and *fats* in large quantities to be "burned" to give the necessary calories for respiration, to keep the entropy wolf away from the door, and about 90 percent is lost to heat in this way. (One gram of carbohydrates converts to about 4 calories while 1 gram of fats gives off about 9 calories.)

In addition the human individual must have *protein* for growth and maintenance. Proteins are both the "building blocks" of the body, making up most of the blood, bones, skin, muscles, glands, and hair, and the enzymes that are responsible for the body's chemical reactions, that is, for life itself. All proteins are manufactured by the body from about 20 *amino acids.* Of these, 8 or 10, the "essential" amino acids, cannot be synthesized internally and must be taken in as food. Plants contain some of the essential amino acids, but no one plant can provide all; soybeans come the closest, lacking only one (methioine) and legumes in general have most of the necessary acids. For those of you interested in a vegetarian diet, this means that a variety of plant foods must be chosen carefully to give a proper diet. Animal protein is by far the best source of the essential amino acids, and it is estimated that between 30 and 40 grams of animal protein daily are needed for normal growth and maintenance. What happens if protein intake is not sufficient? Body growth and functioning suffers, including deterioration of the hair, and in extreme cases leads to *kwashiorkor,* particularly in children who have just been weaned from mother's milk. Kwashiorkor is a protein malnutrition disease marked by no appetite, no growth, vomiting and diarrhea, bloated stomach and limbs, skin disorders, and extreme mental apathy. It can lead to death but more often indirectly causes death by making the individual more susceptible to a host of diseases that would not otherwise be fatal.

Minerals and *vitamins* play important roles in skeletal development and metabolism and cannot be ignored as part of the ecological system in which humans participate. Tables 3.4 and 3.5 list some of the more important minerals and vitamins, their functions in the human body, and their food sources. The importance of minerals and vitamins in understanding human ecology is suggested by the following example. Several years ago the psychiatric anthropologist Anthony F. C. Wallace (1961) proposed that a form of mental illness called "grand hysteria attack" by Sigmund Freud and "pibloktoq" by the Polar Eskimo of northern Greenland has its origin in chronic calcium deficiency. The classic symptoms start with a period of hours or days during which the victim is irritable or withdrawn. Suddenly the victim becomes manic, tearing off clothes,

TABLE 3.4 EXAMPLES OF ESSENTIAL MINERALS AND SOME FOODS THAT CONTAIN THEM

Mineral	Function	Food Source
Calcium	Constituent of bones. Involved in blood clotting and in regulation of nerves and muscles	Milk, cheese, green vegetables, shellfish, hard water
Phosphorus	Involved in bone building, in nerve and muscle regulation, and in the transfer of energy in the cells	Dairy foods, poultry, eggs, and meat
Iron	Component of hemoglobin molecule in red blood cells	Liver, eggs, oysters, molasses, meat, green vegetables
Sulfur	Important for synthesis of several proteins	Meats, fish (high-protein foods)
Iodine	Component of thyroxin molecule in thyroid gland, which regulates growth, development, and metabolic activities	Foods from iodine-rich soils
Zinc	Component of insulin molecule	Green vegetables, peas, beans, organ meats

Source. George Moriber, *Environmental Science* (Boston: Allyn and Bacon, Inc., 1974).

throwing objects, and, in some cases, running. Just as suddenly the victim has convulsive seizures, followed by collapse and then sleep for several hours. The victim wakes up with complete amnesia about the attack and behaves normally. The disorder is, and has been, fairly common among the Polar Eskimos but is presently rare in other parts of the world. However, hysterical attacks of this type were quite common during Freud's time; indeed, it was his work on this mental illness that led to the development of psychoanalysis. Freud believed that the disorder was caused by a traumatic emotional conflict and that it could be cured by "talking" therapy. Disciples of Freud soon applied the psychological explanation to the Arctic variety, pibloktoq, as well. Wallace argues that there is more support for the hypothesis that calcium deficiency is responsible for hysterical attack. The outbreak of hysteria during Freud's time corresponded with outbreaks of rickets and tetany, both calcium-deficiency diseases, in the working classes of the large European cities such as Vienna and Paris. The deficiency was caused by the near absence of calcium in the diet and to a deficiency of vitamin D3, necessary for the body to effectively use what calcium is in the diet. Vitamin D3 deficiency was caused both by improper diet and, because of smog and no access to open spaces, a lack of sunlight (sunlight stimulates the body to produce some of its own vitamin D3). During the early twentieth century the value of sunlight and milk, along with other calcium and vitamin D3-containing foods, was discovered and disseminated among the working classes. The result was astounding—rickets, tetany, and grand hysteria attacks nearly disappeared.

The Polar Eskimo, according to Wallace, live under similar conditions. Calcium in the community of Angmagssalik was always low, and when the fish capelin was not available it fell below the acceptable minimum. Equivalent results have been reported from other communities. Furthermore, the low Arctic sun and heavy clothing combine to nearly prevent the body from manufacturing vitamin D3 and to take the advantage of what little calcium is available. Therefore, if pibliktoq is caused by calcium deficiency, its common occurrence in this region is to be expected.

Once again, however, the reader should beware of concluding that low calcium levels automatically causes hysteria. Wallace points out that the kind of disorder depends upon ". . .various conditioning factors of

TABLE 3.5 EXAMPLES OF ESSENTIAL VITAMINS AND SOME FOODS THAT CONTAIN THEM

Vitamin	Function	Food Source
A	For healthy skin and gums, and night vision	Animal livers, eggs, and dairy foods. Fruits and vegetables contain carotene, which can be converted into vitamin A in the body
B, a group of eight vitamins	For helping to maintain functioning of circulatory and digestive systems	Liver, yeast, whole grains, unpolished rice
C, ascorbic acid	For healthy gums, skin, and blood vessels	Citrus fruits and leafy green vegetables
D	Involved in the absorption and utilization of calcium and phosphorus in the bone	Fish liver oils. Vitamin D can be produced from a precursor in green plants
E	Believed to be involved in reproductive functions	Vegetable oils
K	For normal blood clotting	Green leaves and egg yolk

Source. George Moriber, *Environmental Science* (Boston: Allyn and Bacon, Inc., 1974).

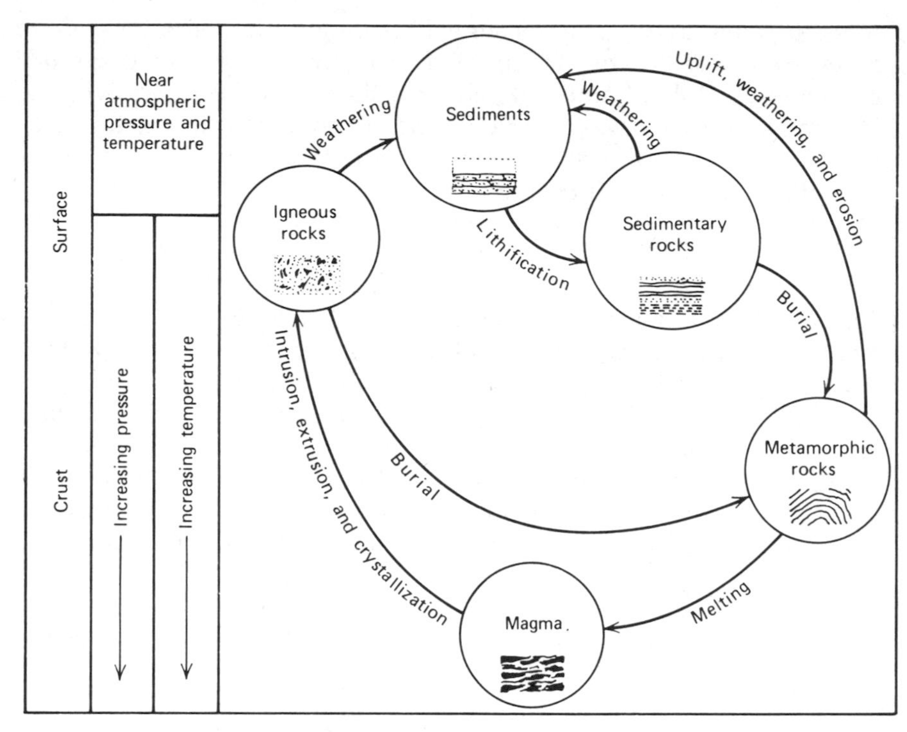

FIGURE 3.4 The rock cycle. (From Joseph M. Moran, Michael D. Morgan, and James H. Wiersma, *An Introduction to Environmental Sciences*, Boston: Little, Brown and Company, 1973. Copyright 1973 by Little, Brown and Company. Reprinted by permission of the publisher.)

situation, personal history, and biochemical individuality" (1961, p. 272). In this case we must suspect that historical "accidents" of personal biology and experiences are amplified into a large, though limited, number of end results. Nevertheless, some individuals, when subjected to the "initial conditions" of low calcium, do manifest the classic symptoms of grand hysteria. Cultural feedback is also important. Wallace feels that the tendency of Eskimos with pibloktoq to run away rather than to become aggressive is a culturally determined personality trait. Other examples include the abandonment of kayak hunting by Eskimo men if their confidence is disturbed, the practice of leaving the community to live a hermit's life, and the willingness of the old to go away and die. "Running away" was not part of the classic grand hysteria of Europe, and the difference is undoubtedly due to cultural variation.

Material Cycles

The complex organic molecules making up carbohydrates, fats, proteins, and vitamins do not occur as free substances on earth and must be synthesized from about 20 basic inorganic materials. Consequently, the origin and movement of such materials as oxygen, nitrogen, hydrogen, calcium, phosphorus, sulfur, and water are important to all ecologists; indeed the concept of the ecological system is partly based upon the

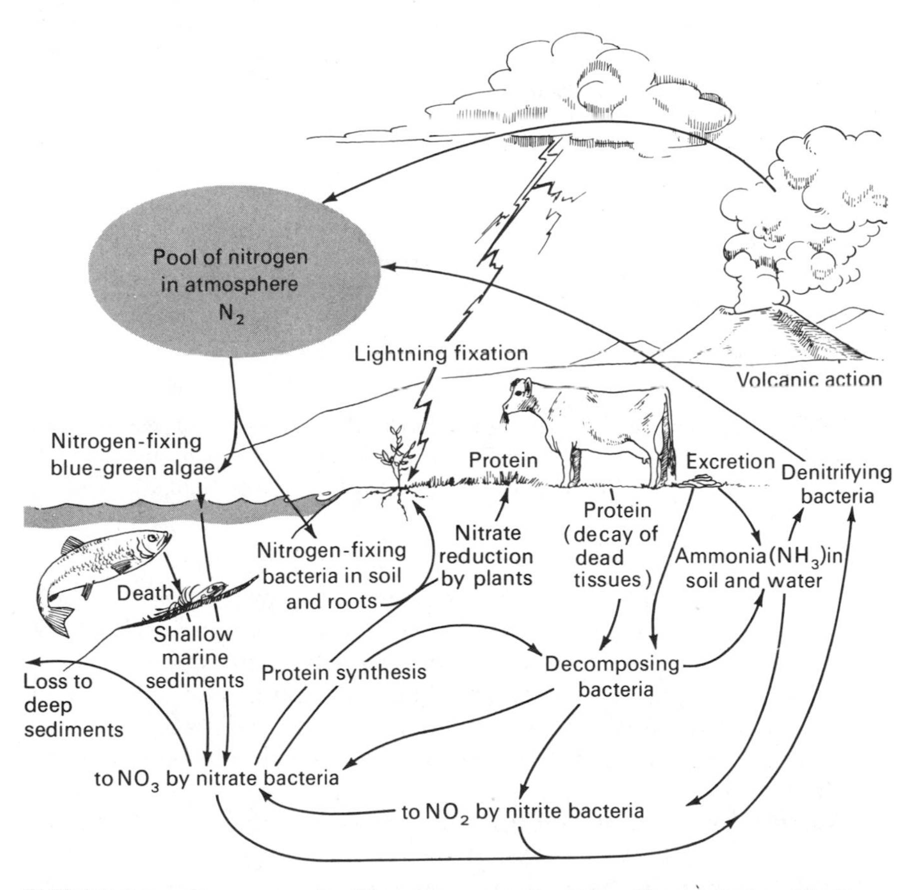

FIGURE 3.5a Gaseous cycles: The nitrogen cycle. (From *Human Ecology: Problems and Solutions* by Paul R. Ehrlich, Anne H. Ehrlich, and John P. Holdren. W. H. Freeman and Company. Copyright © 1973.)

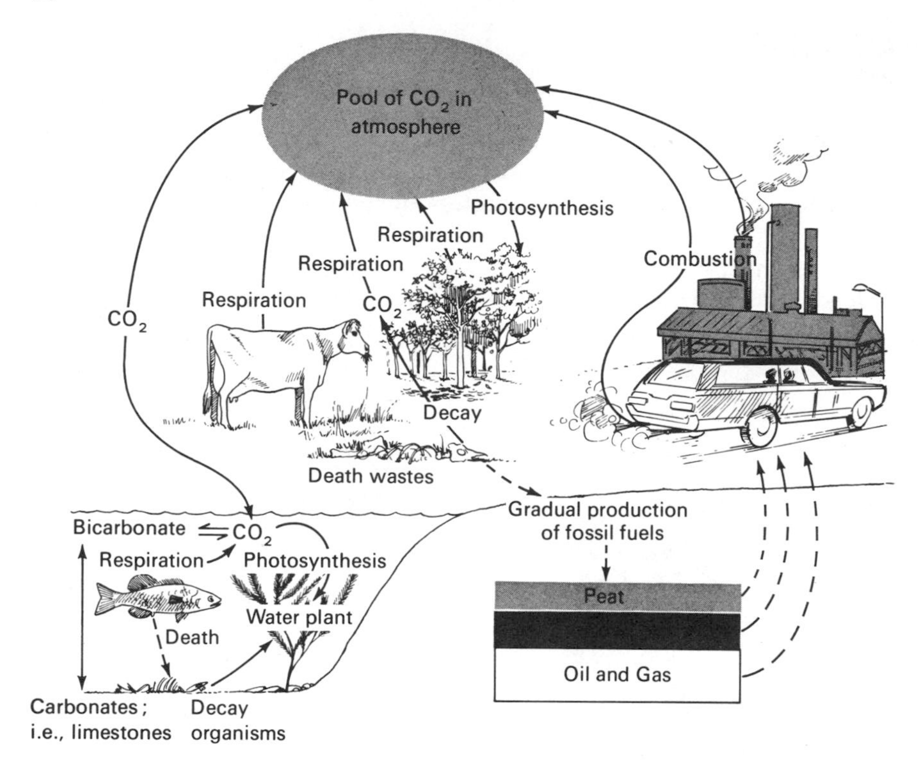

FIGURE 3.5b. Gaseous cycles: The carbon cycle. (From *Human Ecology: Problems and Solutions* by Paul R. Ehrlich, Anne H. Ehrlich, and John P. Holdren. W. H. Freeman and Company. Copyright © 1973.)

existence of *material cycles* that link together living organisms and their inorganic environment. (The following discussion is based upon Kormondy, 1969; E. Odum, 1971.) In theory materials, unlike energy, can be reused time and time again and move in circular paths. Nevertheless, in practice some materials flow so slowly, taking generations if not thousands of years to complete a cycle, that they should be treated as nonrenewable.

Material cycles are made up of *reservoirs* for long-term storage of materials, *exchange or cycling pools* for short-term storage of materials, and a *circulation network* for transporting materials. Let us examine the reservoirs first. Large quantities of materials are stored in the air, or *atmosphere;* the water, or *hydrosphere;* and in the earth's crust, or *lithosphere.* Materials stored in air and water are often in a gaseous state and include such elements as nitrogen, oxygen, carbon dioxide, and hydrogen. Materials in the form of minerals, such as iron, calcium, sodium, phosphorus, and potassium, are most commonly stored in the earth's crust but are also dissolved in water; a few (e.g., sulfur) are found in the air in a gaseous state. Materials stored in these reservoirs have very *low transfer rates* into the other spheres present in ecosystems; that is, materials circulate very slowly.

The exchange pools are also found in the atmosphere, hydrosphere, and lithosphere but are found in addition in living organisms, the *biosphere*. Unlike reservoirs, exchange pools have very high rates of exchange among spheres and are more immediately available.

The transportation routes connecting exchange pools and reservoirs, or circulation networks, are of two types. In many networks the path taken by materials is more or less circular and goes through more than one sphere. Such networks are commonly called *biogeochemical cycles* and, if materials essential to life support are being transported, *nutrient cycles*. Not all materials move from one sphere to another, however, and life may not be included in their pathway. For example, the *rock cycle*, illustrated in Figure 3.4 takes place within the lithosphere alone. Although these cycles

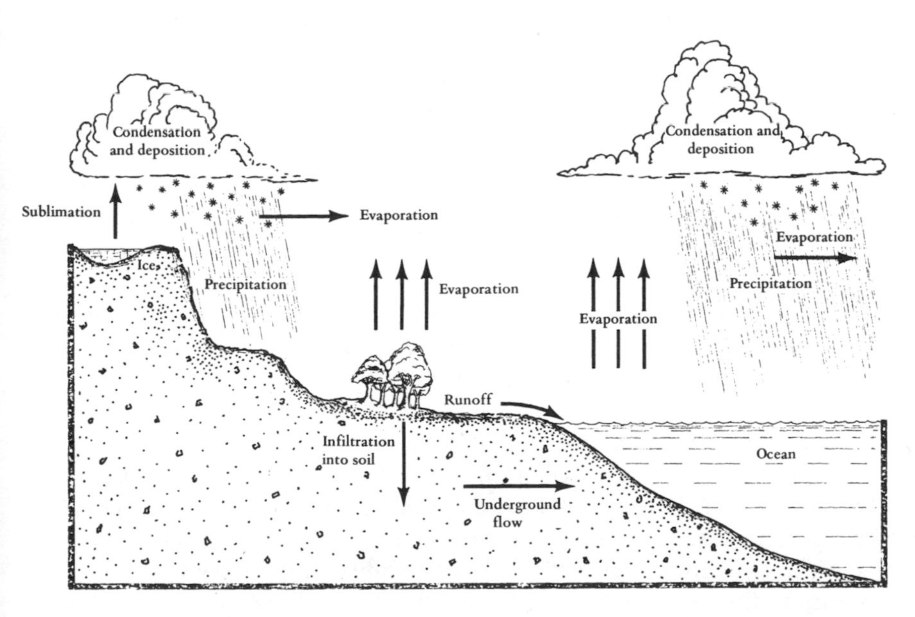

FIGURE 3.5c. Gaseous cycles: The water or hydrologic cycle. (From Joseph M. Moran, Michael D. Morgan, and James H. Wiersma, *An Introduction to Environmental Sciences*, Boston: Little, Brown and Company, 1973 Copyright © 1973 by Little, Brown and Company. Reprinted by permission of the publisher.)

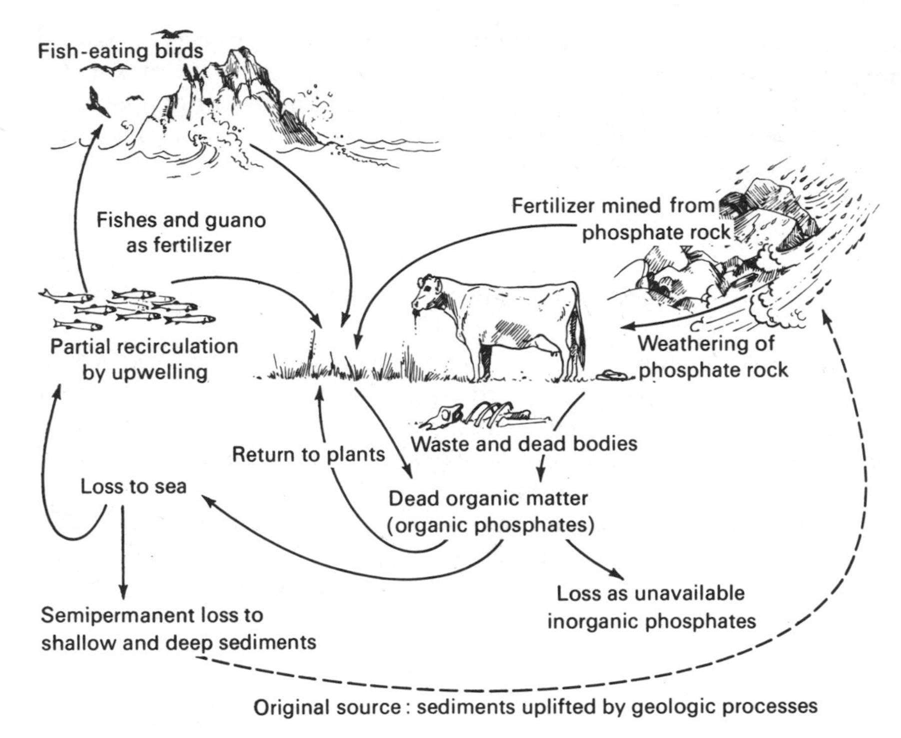

FIGURE 3.6a. Sedimentary or nongaseous cycles: The phosphorus cycle. (From *Human Ecology: Problems and Solutions* by Paul R. Ehrlich, Anne H. Ehrlich, and John P. Holdren, W. H. Freeman and Company. Copyright © 1973.)

do not route materials directly to man, they are of interest to us because of their impact upon human environments. Thus the rock cycle shapes the land surface upon which man must live. Furthermore, the rock cycle is tied into some biogeochemical cycles, such as phosphorus and carbon, which do route materials directly to man.

The precise organization of material cycles varies from one ecological system to another, but on a worldwide scale there are two major types: *gaseous* and *nongaseous* or *sedimentary*. Gaseous cycles have large reservoirs in the atmosphere or hydrosphere, while sedimentary cycles have reservoirs in the lithosphere. Figure 3.5 illustrates the gaseous cycles of carbon, water, and nitrogen, the principal nutrient cycles. The sedimentary cycles of phosphorus and calcium are illustrated in Figure 3.6.

Gaseous cycles are closest to "perfect" because temporary buildups or losses of materials in one part of the cycle can be corrected by adjusting transfer rates. Air and water currents are partially responsible, since local concentrations can be readily dispersed elsewhere, but other factors assist. For instance, a buildup in carbon dioxide can be mitigated by increasing the rate of plant photosynthesis that releases oxygen into the air as a waste product. Sedimentary cycles, on the other hand, are relatively "imperfect." If a transfer rate is disturbed in one part of the cycle (e.g., cultivating fields with an ensuing increase in loss of phosphorus through erosion), transfer rates are not easily adjusted in other parts of the cycle (e.g., the phosphorus is transported to the oceans, where much of it is trapped in deep sediments with *very* slow transfer rates back to land). For all practical purposes the result is a one-way flow of materials.

Like all other creatures, humans ultimately depend upon material cycles to supply their needs. For this reason long-term disruption of the cycles through extensive mining, pollution, and other human activities could be disastrous. The student of human ecology must also realize that materials critical to human life support are not uniformly distributed over

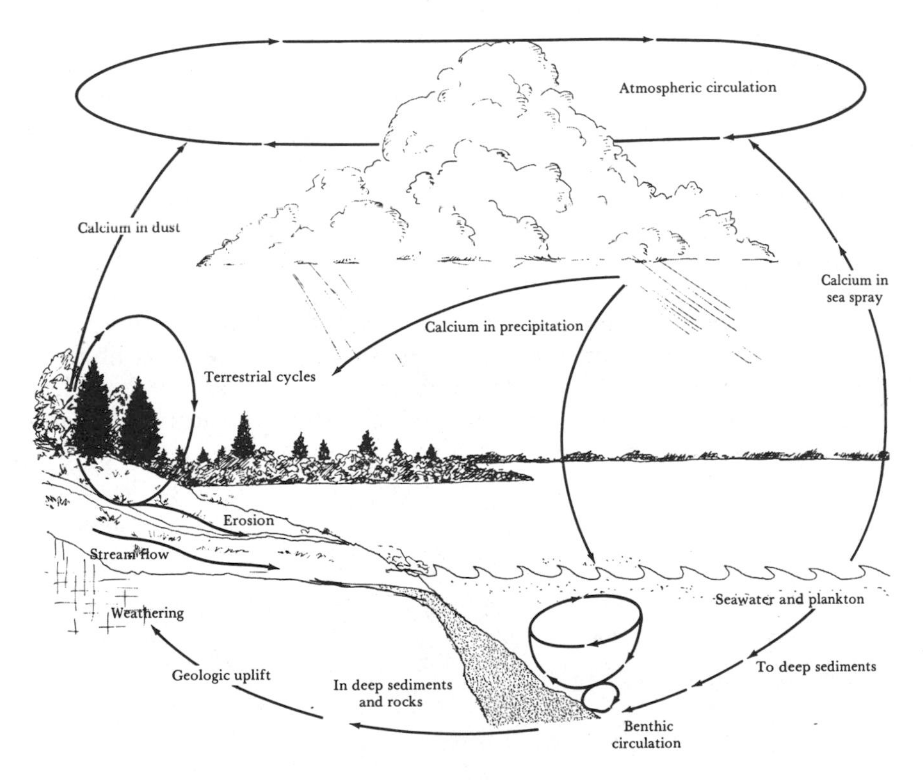

FIGURE 3.6b. Sedimentary or nongaseous cycles: The calcium cycle. (From Robert Leo Smith, *The Ecology of Man,* New York: Harper & Row, 1972. Copyright © 1972 by Robert Leo Smith. Reprinted by permission of the publisher.)

the earth's surface. Some localities are blessed with an overabundance; others are cursed with nothing at all. In the past such problems of geographical distribution have been solved by *trade* and by *coercion,* for example, military conquest. The same solutions are used today. Consequently, the flow of materials in human ecology is as much a *social* process as it is a physical one and problems of distribution usually come out of the social fabric. This is particularly evident in today's world where the industrial nations of the west have monopolized raw materials from all over the globe by a mixture of trade and political coercion.

SUMMARY

Human energetics is the study of how human groups cope with energy problems. Like all other life forms, humans are improbable organizations of matter that must receive regular inputs of energy from their environment. At the same time, energy is being lost in accordance with the laws of thermodynamics. Energetics is the study of the energy exchange that solve, or do not solve, these problems. Feeding, or trophic, exchanges have been the major focus of ecological anthropology, but other exchange mechanisms (e.g., radiation, conduction, and evaporation) must also be considered. Many energy problems and their solutions can be understood by reducing all forms of energy and matter to a common denominator, the calorie. However, humans and other organisms often require energy in a *particular* form, for example, as carbohydrates, fats, proteins, vitamins, or minerals. Limited availability of these forms places severe restrictions on energy exchange. For this reason, detailed information on composition of the diet of human groups is necessary to understand human energetics, as is knowledge of the physical cycles and pathways through which material flow.

FOUR ENERGETICS AND HUMAN SOCIETY

Recognition of a close relationship between energetics and human society goes back at least to Karl Marx. According to Marx, "the mode of production in material life determines the general character of the social, political and spiritual" life (quoted in M. Harris, 1968, p. 232). The ecological approach in anthropology has traditionally accepted such a "materialistic" interpretation of social and cultural behavior. Actually, it is more correct to see the ecological correlates of social organization as not just economic but as *all* environmental problems. Some of those problems are economic, in the sense of the production and distribution of energy and materials, but others are not. Thus social organization can at times be interpreted as a buffer or an information collector/processor to more effectively cope with hazards. Nevertheless, economic problems are often foremost in determining an adaptive strategy; viewing social organization from this perspective can be useful. Two kinds of "economic" organizations stand out. *Organizations of production* directly or indirectly, as a technological superstructure, extract energy and materials from the environment. *Organizations of distribution,* by contrast, define the social network through which energy and materials flow to individuals within the human group.

ORGANIZATIONS OF PRODUCTION

Julian Steward was the first anthropologist to systematically investigate the role of social organizations in extracting energy and materials from the environment. His work on the social organization of the hunters and gatherers in the North American Great Basin (e.g., 1938) led to the definition of several distinct types of organization, each of which represented an adaptation to a problem of subsistence. A *family* organization is adapted to low productivity habitats with widely dispersed and erratic resources. Dispersed nuclear families are diagnostic of this type, with larger aggregations taking place only temporarily when local resources permit. Many examples are found in the arid Great Basin and Plateau regions of North America, specifically the Western Shoshoni. The *patrilineal* band is a multifamily organization, although a small one of no more than 50 persons, that is adapted to collective hunting of small, nonmigratory game animals. In Steward's view the patrilineal band was a rather inflexible organization, marked by distinct territoriality and patrilocality. Both of the latter characteristics he attributed to the fact that hunters in these bands are men, that hunting is improved by having intimate knowledge of an area, and that intimate hunting knowledge can be increased by keeping men in the same region where they were born and raised. *Matrilineal* bands are also multifamily, territorial organizations, but unlike patrilineal bands they are adapted to intensive gathering of abundant plants, for example, seeds or tubers. Since plant gatherers are usually women, Steward believed that the necessity of having intimate knowledge of the area exploited for subsistence would lead to matrilocality and matrilineality. Finally, the *composite* band is adapted to the collective hunting of larger game animals. It is diagnosed by a relatively large group of unrelated nuclear families linked by ritual or fictive kinship ties or by multifamily groups linked by bilateral descent.

Steward's interpretation has been challenged by Elman Service (1971a). Service contends that the patrilineal, patrilocal band had a much more widespread distribution in the past than Steward believed and, indeed, is the prototype organization for foragers. The wide distribution of the patrilocal band is, according to Service, not so much due to adaptation to subsistence requirements as it is due to the requirements of interband relationships, that is, to adaptation to the social environment. Steward's other types are considered by Service to be either anomalous, as in the case of the matrilineal band, or a recent phenomenon caused by contact with Western civilization and subsequent depopulation and resettling, as in the case of the composite band.

However, in 1966 an international symposium on "Man the Hunter" rejected the model of a "universal" patrilocal band and proposed in its place a flexible band organization similar to Steward's composite band (Lee and DeVore, 1968). Members of the symposium contended that recent research among foragers does not support the idea that the patrilocal band is inevitable and that the composite band is merely the product of recent contact, although they recognized that the study of the social organization must consider the effects of contact. In fact, a paper given by L. R. Hiatt (1968) at the symposium gave evidence that there is not a single undisputed example of the patrilocal band in Australia, the area from which Service drew his model. There is also no evidence that groups of males are organized into patrilineal "corporations" for the purpose of protecting rights to a territory. Recent research suggests instead that most foragers move around a lot, in some cases joining different bands from one time of the year to the next and in other cases crossing with ease the territorial "boundaries" of different bands.

Flexibility, based upon bilateral group affiliation, changing group size, and shifting territorial boundaries, was found by the symposium to be most characteristic of modern foragers. This kind of organization is adaptive in habitats with fluctuating resources, such as the marginal regions now occupied by foraging groups. The basic features of the flexible organization were outlined by the members of the symposium as follows:

1. Local groups are small, usually less than 50 persons, and occupy a broadly defined geographical region.

2. Local groups are not isolated social systems but are linked together by innumerable social relationships, including reciprocal visiting, intermarriage, and task group membership.

3. Each group is made up of social equals, and there is no accumulation of either food or personal possessions.

4. Neither land nor resources are considered the exclusive right of local groups, allowing groups to move in order to take advantage of shifting resources; that is, they are noncorporate and nonterritorial.

5. Local groups have an orientation toward bilateral affiliation rather than toward unilineal membership rules.

6. The size of local groups changes as a means of moving populations away from food shortages and toward areas of plentiful food.

Most recently, B. J. Williams (1974) has proposed a model of band organization that attempts to compromise the ideas of Steward, Service, and the "Man the Hunter" symposium. Williams accepts the symposium's argument that the band organization is relatively flexible;

however, he does not believe that composite bands and bilateral affiliation are typical. Rather, he accepts Service's argument that the patrilocal band was nearly universal in the past but feels that the organization is much more flexible than either Service or Steward envisaged. The basic assumptions of the model are as follow:

1. Local groups of foragers hold economic territories.

2. The local groups having the most social autonomy are bands of about 50 persons each.

3. Newly-weds tend to remain within the territory of the husband's band; therefore, the band is both patrilocal and patrilineal.

4. The autonomy of local bands is decreased by a rule of exogamy; that is, bands are required to intermarry.

5. Bands that are affiliated through intermarriage are also linked by matrilateral, primary kinship ties.

6. Bands closely affiliated through marriage ties make up a "connubium," having an average size of about 500 persons.

7. Marriage-affiliated bands and the connubium cooperate in task groups, the use of "windfall" resources, ceremonies, and in moving band members from territories with temporary food shortages to territories with more abundant resources.

Of course, the reader must constantly keep in mind that these models of flexible band organizations are based upon observations of foragers occupying marginal habitats, habitats that are marked by shifting resources and/or a low carrying capacity. It seems doubtful that this has always been the case, at least not until the expansion of farmers into the more favorable habitats. Therefore, the archaeologist particularly must be wary about applying the flexible model wholesale to the study of past foragers.

Tribal Organizations

The concept of band organization given above is based upon flexibility and allows for alternate dispersion and aggregation of local groups as a means of exploiting resources. Kinship ties through proscribed intermarriage (that is, exogamy rules) make this possible, as do nonkinship ties such as ritual and voluntary association. In effect the band connubium as an ecological entity is often no different than the *tribe;* at the very least the tribe, as classically conceived, is no more than one end of a continuum of band organizations. According to Marshall Sahlins (1966) and Elman Service (1971a), who have given the most recent and refined definition of the tribe, tribes are collections of interrelated bands with some *emergent* features and exist for the purpose of controlling competition for scarce resources, particularly land made valuable by the advent of farming strategies. Local groups are tied together by tribal organizations in much the same way as the band connubium; that is, kinship ties through intermarriage are fundamental but also include a number of "pantribal" integrating mechanisms such as secret societies, age grades, and various ritual institutions. Social control in tribes, like bands, is not based upon specialized social positions and roles but upon social positions and roles that already exist, those immersed with the structure of kin relations. Sahlins and Service recognize the similarity between bands and tribes but argue that the tribe is different because it has a stronger sense of community and distinct territorial boundaries, both in keeping with their idea of the tribe as a means of controlling scarce resources. However, Fried (1967) contends that ethnographically documented tribes often do not have distinct boundaries and membership. For example, Fried cites evidence that the Chimbu, highland farmers of New Guinea, were, prior to European contact, made up of tribes whose segments shifted membership and land boundaries from time to time. Roy Rappaport (1968) gives data for the Tsembaga Maring tribes suggesting the same kind of instability. Furthermore, Fried argues that the strong ethnic identity of tribes is a recent adaptation to social environments dominated by colonial powers. That is, tribal organizations are responses to alien intruders ("social environment"), not to the need for controlling resources.

Ranked and Stratified Organizations

Modern nation-states are often considered to be the epitome of production organizations, and there can be little question that the massive agricultural and mining projects common to such polities are the largest yet known to man. Indeed the very existence of industrial states depends upon their ability to effectively exploit raw materials on a large scale. Territorial conquests, colonialism, and widespread trading networks are but a few of the ways to achieve this end. Similar strategies have been used by archaic states in the past. For example, the Aztec state controlled mining activities in northern Mexico, and the Inca state used public *corvee* labor for the construction of terraces, irrigation works, storehouses, and other farming projects. The vast irrigation projects of dynastic Mesopotamia are also typical, as are the state-controlled trading networks of the Aegean, Mesopotamian, and Mesoamerican states. The ability to organize large groups for exploitation is, in the case of states, based upon the *legitimate* use of force. That is, the elite groups representing the state are granted *authority* by their constituents that can be used for organizational purposes, organizations that are used not only to provide life support for the people of the state but also to provide a rationale for the existence of the elite power.

Production organizations are also found, although smaller and less common, in societies in which access to authority and prestige is unequal but in which high status does not confer the use of legalized force. Marshall Sahlins and Elman Service refer to this kind of society as a *chiefdom,* and Morton Fried includes it within his *ranked* society. Chiefdoms are marked by a system of hierarchical ranking of a person within the context of kinship. A single individual, the chief, occupies the apex of the rank "pyramid" and has the highest prestige and greatest authority. Although all persons within the society are considered to be related in progressively lesser degrees to the same ancestor as the chief, most of the prestige and authority are vested in a small group of closely related kinsmen. The unequal social system is used as a fulcrum to organize task groups for some public projects and, as we shall see shortly, as a means of redistributing goods and services. The authority of chiefs and elite groups is, however, based upon ritual and "conspicuous consumption" rather than upon legitimacy. Consequently, severe limitations are placed upon the chiefdom as an organization of exploitation, since the "mythological charter" of the chief is fragile and often does not survive even until his death.

In many parts of Polynesia and elsewhere social ranking is manifested in lower-level social organizations. The occurrence of ranked families in Fiji, for example, is believed by Marshall Sahlins (1957) to be related to the necessity for coordinating task groups at the local level and for redistributing resources in a technoeconomic system based upon *dispersed* farming land. Dispersion creates problems of travel time between the fields and the settlement so that a small nuclear family could not take care of the fields and still maintain a sedentary existence. An extended family, however, has enough persons to allow some to spend a few days away in the fields while still keeping at home a labor pool sufficiently large to handle the requirements of a sedentary community. Sahlins observes that this is facilitated by dividing the extended family into cooperative labor groups defined along sexual lines. Thus the females of the family join forces to do housework, care for children, and gather resources (such as fish) that are close to the homestead. Males, on the other hand, are the farmers of the distant fields, assisting at home primarily by building houses when needed and by doing some fishing. Dispersed farming land also means that a variety of crops can be grown, taking advantage of differences in elevation, soil, rainfall, and so forth. However, because of the many different activities required to take care of specialized fields, a relatively large labor pool must be available along with some means of dividing up and coordinating the various tasks.

A nuclear family has neither the size nor the organization required but the Fijian extended family does. Division of labor and the coordination of activities are made possible by a system of ranking based upon kin terms and a complex system of etiquette. Rank is assigned by generation and birth order. The father of the family has the highest rank and the most authority, followed by his children in order of birth. Distinctions of seniority are found in kin terms, and the expected behavior of senior and junior members of the family toward each other is similar to that between chiefs and commoners at higher social levels. One of the more important economic manifestations is that seniors can demand and receive goods and service from juniors at any time; seniors are also expected to give assistance to juniors when needed but only after appropriate demonstrations of humility. The seniority system controls both subsistence activities and their products. Thus the senior male or female (the wife with most senior husband) directs the labor force and assigns tasks. Although each mature man "owns" a yam garden, taro patches, and other fields, the garden products are not his to do with as he sees fit; rather, they are controlled and distributed by the family head, that is, the most senior male. As a result the resources coming from diverse fields can be pooled into a common pot and used by the entire extended family.

ORGANIZATIONS OF DISTRIBUTION

Even from the ecological perspective, social organization cannot be understood entirely as a way of exploiting energy and materials. Social and cultural behavior also link people together into organizations that distribute energy and materials from the producers to individual consumers. The variety of such organizations is immense but those related to reciprocity, redistribution, and markets are the most important (Polanyi, 1957).

Reciprocity

Generalized reciprocity is the predominant form of economic distribution under conditions "in which the stimulation of intensive extra productive effort would have an adverse effect upon group survival" (M. Harris, 1974a, p. 126). These conditions are commonly found in ecological systems participated in by foragers. Low productivity and a delicate balance between man and nature place stringent limitations upon exploitation efficiency and intensity. A *low* and *irregular work effort is* necessary to maintain those limits; Richard Lee's (1968) work among the Dobe !Kung Bushman shows that they are among the most leisured people on earth. Furthermore, the Bushman and other generalized foragers make sure that no conditions exist for "social climbing," since competition for social status seems inevitably to lead to more work on the part of "achievers" exploitation, exploitation that if uncontrolled can destroy the group's life support. According to Marvin Harris (1971, p. 243) the ethic of generalized reciprocity is one way of preventing excessive exploitation because it is both the cause and the effect of a low and irregular work effort.

In a general sense reciprocity of this kind seems to be "whimsical" gift giving; that is, gifts are given whenever the giver feels like it and with no expectation of return. A similar form of exchange is found in the United States among close members of a family and probably exists universally. However, only among generalized foragers is it extended to the entire society and to almost all economic transactions. The "whimsical" overtones are more apparent than real, and Harris lists the following essential features of the reciprocity ethic (1971, p. 238):

1. Gift giving with no *immediate* expectation of return.

2. No attempt to assess the value of gifts.

3. No attempt to make gift giving "balance out"; that is, records, mental or otherwise, are not kept.

All of these exist to assure that no person boasts about his generosity in gift giving and creates a social atmosphere conducive to status competition, therefore stimulating extra work and intensive resource exploitation.

Reciprocity is not always of the "generalized" form; *balanced reciprocity* and *negative reciprocity* are two other related kinds of economic distribution. Balanced reciprocity takes place between persons who are not close family but are social equals and have some kind of *personal* relationship. That is, the exchange partners may be distant relatives, members of the same voluntary association, or schoolmates. Balanced exchange involves not only an expectation of return, although not necessarily an immediate one, but also some accounting of the value of the gift. Negative reciprocity carries this tendency to its logical extreme. Exchange is usually immediate and strict accounting of the value of the products of exchange is kept. The partners in negative exchanges are social equals, as in generalized and balanced reciprocity, but are not immersed in a network of personal relationships. That is, the exchange is impersonal.

Social environment influences the form of exchange that takes place, as well as physical environment, and changes in the social environment is a frequent cause of shifts in the mode of exchange. Napoleon Chagnon (1968) gives data from the Yanomamo farmers of Brazil and Venezuela to illustrate the role of social environment fluctuations. Hostile Yanomamo villages reestablish "diplomatic" relations and become political allies through a progressive series of steps marked by shifts in the form of economic exchange taking place between the villages. The first step is dominated by negative reciprocity in which the villages suspend hostilities and exchange the products in which each village specializes, for example, pottery or hammocks. Gift giving at this stage is strictly on a one-for-one basis, and the exchange is immediate. After a suitable period of time, the exchange relationship is escalated to balanced reciprocity. Feasts are given to which the exchange partners are invited and gifts bestowed. These gifts need not be returned until later, when the recipient gives a "return" feast. Political alliances between the villages are now fairly substantial but are further strengthened by intermarriage, the third step. Intermarriage often takes the form of men from allied villages exchanging unmarried sisters for purposes of marriage. The consequent establishment of kinship ties escalates the exchange relationship to generalized reciprocity. Gifts are given and received frequently and the political alliance between the villages is at its peak. In the event that the alliance breaks down, exchange of any kind is ended abruptly and hostilities resume.

Redistribution

In societies marked by social inequality, goods and services often flow from persons or groups low in prestige and authority to persons at the apex of the social pyramid and back again. This form of economic distribution is called *redistribution* and usually takes place between relatively large social groups, for example, the village or lineage (Fried, 1967, p. 117). One of the more intriguing strategies of redistribution is the elaborate and massive "display" feasts given by social "climbers" for the purpose of gaining status. The best known example is the *potlatch* practiced historically by the Kwakiutl of the American Northwest Coast. Potlatching was a way of gaining social status by giving huge feasts to which status rivals were invited and massive amounts of food, blankets, clothing, and other prestige items either given away or destroyed. The more spectacular the gift giving or destruction, the more rivals were shamed and the more status achieved. A similar example is the "big man" system of New Guinea and Melanesia. Social status is again gained by accumulating as much food as possible, through relatives, friends, or begging, and giving a feast at which the food is given away to the guests. Obviously, "who one knows" is crucial; the more relatives and friends one has, the larger the feast that can be given and the more the status gained. The system is also marked by a positve feedback loop, since high status attracts followers and a large following makes it possible to give larger feasts. Under these circumstances, of course, a big man must work continually in order to maintain his status; otherwise, rivals will soon overtake him. Consequently, the achieving person must organize family and friend to help in farming, fishing, and other means of food production and must work much more than would appear to be necessary.

Like other redistributive strategies, competitive feasting is adapted to ecological systems that can withstand intense exploitation for long periods of time. According to Marvin Harris it is also associated with the following conditions (1974a, pp. 118–119):

1. Everyone has equal access to the means of production.

2. There are no political organizations to tie together local groups into a single economic network.

3. Local groups occupy environments that have annual fluctuations in productivity, fluctuations that are often of the "boom-bust" sort.

Competitive feasting under these circumstances keeps people working hard enough to assure that a food surplus in one area will be available for redistribution to offset a crop failure elsewhere. Potlatching is illustrative.

The Kwakiutl were organized into localized kinship groups called *numaym*. (The following discussion is based upon Drucker, 1963; Piddocke, 1965.) Membership into the groups was by bilateral affiliation and was associated with a mythological place of origin. The *numaym* was a corporate entity, owning fishing locations, hunting territories, houses in the winter village, and other microenvironments and tangible goods. Leadership was vested in a "chief," who was the custodian of the resources and goods of the *numaym* and who also performed rituals. As long as the microenvironments of the kin groups were productive, the organization was self-sufficient and not tied to other such groups. However, the fishing stations and hunting localities did fail at times. Slight shifts in the ocean current, for example, were enough to change a salmon run from one part of the coast to another and, therefore, from the microenvironment of one *numaym* to that of another. When the resources of a *numaym* did fail, warfare was sometimes the result; more often, however, surplus food from other *numaym* could be exchanged for material wealth, such as blankets, slaves, boxes, and other prestige items. As suggested, competitive feasting, in this case the potlatch, is a way of assuring that a surplus will be available. Also, if the famine continues and the material wealth of a *numaym* is exhausted, the potlatch is a way of getting back some of the wealth in exchange for social status.

Another typical organization of redistribution is illustrated by the *ramage,* common in parts of Polynesia. Although the exact form of the ramage is rather rare, the hierarchical framework of unequal social positions that it provides is a common way of distributing goods and services. Let us consider the ramage system on the island of Tonga as an example. (The following discussion is based upon Burrows, 1939; Sahlins, 1958.) Tongan society was based upon a unilineal descent group with no exogamous or endogamous restrictions, called a *haa* or lineage. Raymond Firth, in a study of Tikopia social organization, has referred to this type of unit as a ramage. Relationships in Tonga are generally traced in the male line but are frequently alternated with female descent for purposes of regrouping. Each ramage is stratified on the basis of "distance from the senior line of descent from the common ancestor" (Sahlin, 1958, p. 140) and contains two or three status groups or levels. The top status level of all ramages is that of the chiefs and their immediate relatives, including the heads of major ramages and the three paramount chiefs—Tui Tonga, Tui Haa Takalaua, and Tui Kanokupola. Paramount and other chiefly positions were based upon the inherited power of office rather than upon achievement. Inherited *mana,* or disembodied power, is the main source of the political position but, in the case of the Tui Tonga, direct divine descent underlies the position. The tendency for paramount chiefs to choose immediate relatives to rule provincial regions often proved costly in this regard, since *mana* could be used as a justification for revolts led by these officials. Contrary to state organizations, however, Tongan chiefs had no recourse to legal force, no armies with which to enforce their will. Rather, like most of the African kingdoms, the power of office was based upon mythological charter, the general belief that *mana* and divine attributes were part of holding office. A second level of status was that of the chief's attendants, while a third was comprised by commoners. In some cases a fourth level could be distinguished, that of war prisoners or slaves. The largest group of individuals in each ramage was made up by commoners.

Ramages were arranged according to the relationship of heads to a common ancestor. All major ramage heads, however, were relatively equal in "descent distance" so that ranking of ramages was not pronounced. Major ramages were split into smaller ramages, the smallest of which corresponded to the patrilocal extended family. Small ramages were frequently produced by a process of fission and were often accompanied by movement to a new area. All lower-level ramages were obligated to higher-level ramages of which they were a part, particularly to the head of the major ramage.

Tongan ramages were correlated with territorial divisions. Edwin Burrows (1939) has referred to the relationship of Tongan kinship units (ramages) to territorial divisions as the "intermingling of breed and border" to indicate the presence of several kin groups within a single territorial unit. However, a more precise statement would be that low-level ramages occupied diverse areas within a territory owned by a higher-level ramage. Human settlements were dispersed in accordance with the distribution of natural and artificial resources. The ramage system facilitated the control of these resources by a system of economic redistribution. Hamlets or households were managers of small tracts of land or marine resources, upon which a few resources were exploited. Households were grouped into low-level ramages, equivalent to the extended family, with corresponding "pooling" of resource areas. In turn low-level ramages and their territories were grouped into higher-level ramages and "managed" by heads of the larger groups. Each managerial unit and associated manager was subordinate to a higher authority, up to the paramount chief. Resources from local areas were allocated along ramage lines and then redistributed as follows:

Surpluses of food and other resources are constantly or periodically, depending on the crop or time of production, passed up the social hierarchy from household heads to heads of larger ramages and often to the paramount chief himself. Heads of ramages thus form the apex of an organization of accumulation. The paramount chief is the apex of an accumulation system of the entire society. The collected goods are then periodically redistributed among producers. The periods of redistribution and often of accumulation, are occasioned by the many large religious ceremonies and feasts, life-crisis rites, intertribal visits, and the like. The scale of collection and redistribution correlates with the status of the presiding chief. For example, the entire society usually attends and contributes and receives food and goods if the life crises are in the high chief's family. A smaller gathering will mark a similar occasion in a family of lesser rank. It is for purposes of accumulating the wherewithal to celebrate such events that stewards exercise their tabu powers with respect to the resources of their areas.[1]

MARKETS. Market distribution of goods and services is most familiar to the western world. Buyers and sellers in a market arena are expected to enter into economic transactions on a purely *impersonal* basis, immediately removing the market economy from the kinship and prestige spheres of reciprocity and redistribution. Buyers attempt to purchase in the market the best products for the least cost, while sellers attempt to maximize their profits, that is, to increase the difference between what they pay for the product and what they sell it for. A final selling price is negotiated by "shopping around" or "higgling-haggling." The selling price in a market is dominated by the supply of the product and its demand. Thus if the supply is large and demand low, the selling price is low; however, if the supply is small and the demand great, the selling price is high.

In the classic market system of the Western world, any impact that the manufacture or sale of a product may have outside the market arena is an *externality* and is not considered as part of the price. For example, the discharge of waste into the air as part of the manufacturing process is an externality because no cost is incurred. Obviously, however, if the waste is thereby deposited on a neighbor's clean laundry hanging in the backyard and the neighbor must pay to have the laundry cleaned, a cost *has* been incurred although paid for by someone else. The whole problem of externalities is an important part of present-day concerns for environmental protection.

SOCIAL ORGANIZATION AND PRESTIGE

Economic anthropologists have traditionally viewed economic exchange as more than just a way to produce and distribute energy and

[1]From Marshall Sahlins, *Social Stratification in Polynesia,* Seattle: University of Washington Press, 1958, p. 149. Copyright © 1958 by the University of Washington Press. Reprinted by permission.

materials. The exchange is deeply rooted in a web of social relationships and may exist independently of material wants and needs. In effect economic exchange can be completely understood only in terms of assigned social value. One need only point to the numerous systems of exchange involving only "prestige" items, not consumable food or materials, in support of such an argument. There is no question that more or less arbitrarily assigned social value is an important rationale for the existence of many kinds of economic exchange. At the same time it is not always true that the exchange of prestige items has no relevance to ecological relationships. Mervyn Meggitt (1972) gives an example from the Mae Enga of the western highlands of New Guinea. They are shifting farmers organized into named and localized particlans that are exogamous, politically independent and economically self-sufficient. In turn the patricians are subdivided into subclans and patrilineages. There are no hereditary or elected political leaders, but there are "Big Men" who achieve power during their lifetime through social manipulations. Big Men are leaders of clans and other kinship groups and it is among these men that the Te exchange cycle is centered. In the Te cycle gifts ranked in prestige are presented at ceremonial occasions for various social events or infractions, including homicide, death, illness, marriage, insults, and broken betrothals, or they are simply given as gifts to create an obligation from another person. The reasons for gift giving are ranked in importance and the rank of the appropriate gift is correlated with the reasons. The appropriate gift is mediated by the Big Men and the object of Te exchange is to accumulate as many gifts from as many different persons as possible. Of particular importance is building up *debts* from other persons because these individuals are then obligated to give assistance when needed.

The Te cycle starts with the presentation of initiatory gifts to one or more members of a neighboring clan. The presentation is made publicly by the gift-givers clan and its Big Men. After receiving these gifts, the neighboring clan initiates gift-giving of its own to its neighbor, and so on down the the line *in one direction.* According to Meggitt it takes two or three years to reach the end of a line of Mae clans. After a time the clan that first initiated the gift-giving demands of its neighbor a repayment in pigs, the most prestigious gift, and the demand is relayed in turn down the line to the clan at the other end. If the members of the distant clan feel that they have accumulated enough pigs to begin repayment, they hold a public ceremony during which pigs are repaid to their neighbor, setting in motion a series of repayments back down the line to the original gift-giver. This takes six to nine months. Finally, the end clan that began the repayment series demands of its neighbor a return gift, and this demand is circulated down the line to the other end. This clan, the last to receive pigs and the original giver of initiatory gifts, then butchers about half of its pigs and begins a series of return gifts of pork sides. A complete set of transactions involving return gifts also takes six to nine months so that the entire Te cycle takes about four years. The cycle is restarted when the clan to begin return gift-giving demands new initiatory gifts, although this time the directions in which the gifts travel are reversed from the original cycle.

Meggitt believes that the Te exchange cycle is tied to ecological processes maintaining a balance between Mae population size and available farming land. The reason is the impact of the exchange system upon the fortunes of individual clans in warfare. The patriclans are the main antagonists and the process of warfare is typically induced by a clan's outgrowing its farming land. However, expansion into adjacent land can occur only at the expense of a neighboring clan and only after hostilities have escalated into a "rout," routs taking place whenever the antagonists are mismatched in military prowess. It is at this point that the Te exchange cycle becomes significant, since the success of the antagonists in obtaining and holding political allies is crucial to success in war. Among the Mae, political alliances are determined by the *prestige* of the clan, prestige which is in part based upon the number and rank of gifts received in the Te exchange cycle and upon the number of other clans owing gifts to them. Big Men are often instrumental in the prestige of a clan, and therefore in its political alliances, because of their role in negotiating gift exchanges. Demographic size is another important factor because, other things being equal, a large clan tends to be more prestigious than a small one. However, both are interrelated, since the fortunes of a Big Man are generally dependent upon the prestige of a clan and prestige is at least partially due to size. Whatever the actual reasons for prestige, prestigious clans attract more political allies, are more likely to be the victor in a "rout," and are, therefore, more likely to expand into the land once held by neighboring clans. By contrast clans that are not prestigious tend to lose political allies, to lose in warfare, and to be decimated or driven out of the Mae territory. Since the Mae territory does not have any vacant land, any expansion of a Mae clan must result in a simultaneous contraction of the land of another Mae clan. The interaction between expansion and contraction of clans is an effective regulator of the balance between population and land.

SUMMARY

Economics and ecology converge in the study of human energetics. The relationship is easier to grasp if the human individual is made the focal point. An individual copes with the problems of access to energy and materials through a social and cultural matrix, not independently. The matrix provides ways of cooperating with other individuals to jointly exploit an energy or material source and to receive a share. Economics is the study of those ways. At the same time the individual, by virtue of economic ties, becomes immersed in a network of relationships involving humans, their energy and material sources, and the living and nonliving environment that affects all of them. Ecology is the study of those relationships. There is simply no hard and fast line that can be drawn between the two.

FIVE
ENERGETICS AND HUMAN INTERFERENCE

Interference with food chains and other ecological relationships is perhaps the most distinctive characteristic of the human species. Disruption, insult, shock, and subtle manipulation are the keys to human success, the pathway that has led to man's present dominance on earth. Again this means that an equilibrium view of the ecological system is not suitable for the study of human ecology. In Chapter 3 several ways in which humans manipulate, change, and create feeding relationships to improve their share of energy and nutrients were mentioned. A more detailed discussion is given here. Briefly, four types of interference are considered:

1. Manipulating or creating "pulses" of energy or nutrients.

2. Increasing net production by manipulating the process of ecological succession.

3. Increasing gross production by removing physical limiting factors.

4. Manipulating the structure of food organisms so that a greater proportion of its biomass is eatable or otherwise usable by humans.

MANIPULATING PULSES

Many ecological systems in which humans participate are dominated by periodic "surges" or "pulses" of energy and/or materials, controlled on the one hand by nature, such as the seasonal flooding of river valleys, and on the other hand by culture, such as the artificial flooding of rice paddies. Humans have been exceptionally successful in such systems by diverting and commandeering the pulses.

Natural Pulses

Betty Meggers (1971) describes the floodplain of the Amazon River, or *varzea,* as having an annual cycle controlled by the rise and fall of the river itself rather than by local climatic factors. The Amazon is fed by tributaries from the northern hemisphere that usually reach their peak height between April and August and by tributaries from the southern hemisphere that reach their peak between October and April. Interaction between the tributaries results in a single peak for the Amazon, usually around June, during which time the *varzea* is flooded. Because of several major tributaries flowing from the young and soilrich Andes, the so-called "white water" rivers, the Amazon receives a tremendous load of silt, which it deposits upon the *varzea* during periods of flooding. The annual silt load creates a pulse of materials into the *varzea* system that has a number of ecological consequences. The foremost is *diversity.* Numerous lakes, ponds, and channels cover the floodplain because of the cutting and sculpturing action of the river; each of these comprises an ecological microcosm and each has somewhat different characteristics. Thus, at any given time, the *varzea* has a remarkable variety of plants, animals, and general ecological conditions. The impact of annual cycles in local rainfall causes temporal diversity as well. During the period of maximum rainfall the mosaic of lakes, ponds, and channels receives surface runoff of water with relatively little material content. Prior to the flooding of the Amazon, therefore, many small bodies of water are filled with nearly pure water.

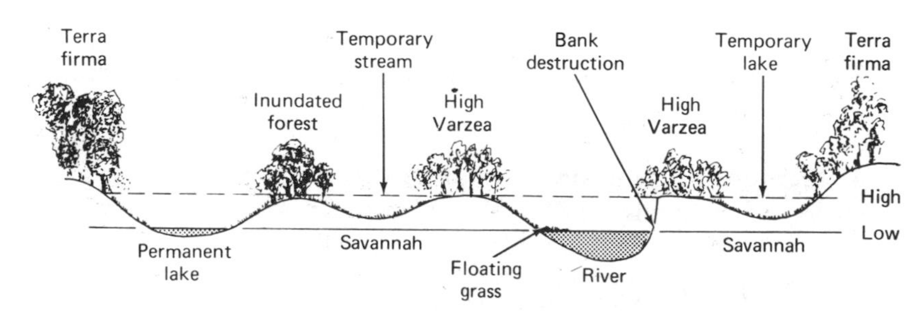

FIGURE 5.1 The Amazon varzea. (Reprinted by permission from Betty J. Meggers, *Amazonia,* Chicago: Aldine Publishing Company, Copyright © 1971 by Betty J. Meggers.)

However, as the Amazon begins to rise, a process requiring nearly eight months and an extraordinarily long time in comparison to other major rivers such as the Nile, the spreading flood waters, carrying a large load of silt, penetrate into, and in many cases completely cover, the floodplain lakes. The result is continuing change in the composition of the *varzea*. Figure 5.1 illustrates an idealized cross section of the *varzea* showing the ecological mosaic and its fluctuation during periods of high and low water.

How do humans participate in the *varzea?* Traditionally, an important role was created by using farming methods. The annual input of materials from flooding provides an ideal foundation for intensive farming without the necessity of using artificial fertilizers to maintain soil fertility. However, the very process by which the all-important nutrients are transported to the fields, flooding, imposed a regime on farming that is reflected in the organization of human societies occupying the *varzea*. During periods of high water not only is farming impossible, but also wild plants are submerged and unavailable and aquatic animals are widely dispersed and hard to catch. However, during low water periods farming is possible and a vast variety of wild plants and animals suddenly becomes available, often in great abundance. The *scheduling* of human activities during low water is therefore critical, as is *storage* of foodstuffs and later *allocation*. Meggers describes how this is done:

During the period of low water, all foods are simultaneously at peak abundance and must be gathered in sufficient quantity not only to satisfy immediate daily needs but to accumulate a surplus for consumption during the months of reduced productivity. The most efficient way to cope with such a situation is to split the labor force and allocate to each group a specific type of activity. Furthermore, agricultural operations on the varzea *must be systematically organized because of the time limitations imposed by the regime of the river. If planting is delayed too long, the crop will not mature before inundation; if planting is done too soon, seeds or cuttings may rot before they sprout. Timing is even more critical when two harvests are programmed, as they were in the case of maize. The expertise needed to manage* varzea *agriculture efficiently would almost inevitably have led to the emergence of specialists, who could direct part of the work force while other members of the community concentrated on hunting, fishing, and gathering activities. Some form of labor allocation is not only a logical response to the pattern of seasonal abundance but is also compatible with the degree of social stratification evidenced by the Omagua and Tapajos. Both were ruled by a high chief who had the power not only to issue orders but to command obedience.*"[1]

[1]Reprinted by permission from Betty J. Meggers *Amazonia* (Chicago: Aldine Publishing Company), pp. 141–142. Copyright © 1971 by Betty J. Meggers.

The surge of energy required to take advantage of pulses is often based on the diversification of assigned tasks and their organization under a designated or hereditary authority, as in the societies occupying the *varzea*. In other cases, however, the use of domesticated animals and machines powered by fossil fuels is an equally common and often complementary solution. Large domesticated animals, such as the "sacred" cattle in India, sometimes appear to be a useless drain upon available energy, particularly since they are used for human purposes only a few days or weeks out of the year. However, living within pulsating systems requires close attention to timing and scheduling of activities and the coordination of all available sources of energy at precise moments; energy can be stored in large animals and much of it used at one time in the form of rapid plowing, planting, cultivating, and harvesting, thus providing the surge needed to effectively release a pulse for human use.

Created Pulses

The intensive cultivation of wet rice in southeast Asia is based upon the maintenance of an artifical ecological system into which are injected pulses of water and nutrients. Wet rice is planted in a dyked or terraced field into which water is allowed to flow. Water is usually taken from a river or stream and diverted into the field by means of a water shovel, wheel, or pump and a series of canals. Along with the water comes a load of sediment derived from soil erosion in the river's drainage basin. Once injected into the field, the water-sediment mixture begins to ferment into a "brew," in which nitrogen-fixing blue-green algae thrive and to which rotting remains of harvested plants are added (Geertz, 1963 pp. 29–30). Like any "brew" rice fields begin to acquire their own special characteristics, more or less independent of the silts and fertilizers added to the field. Rice seeds are planted in a few especially prepared "nursery" fields and allowed to grow until they are old enough to transplant. At this point they are removed to the fields and transplanted in such a way that they are evenly spaced, far enough apart to prevent competition between individual plants but close enough together to provide mutual protection from being blown into the water by a strong wind (Hanks, 1972; p. 38). From this point on the control of water flowing through the system is crucial:

Excessive flooding is often as great a threat as insufficient inundation; drainage is frequently a more intractable problem than irrigation. Not merely the gross quantity of water, but its quality, in terms of the fertilizing substances it contains (and thus the source from which it comes) is a crucial variable in determining productivity. Timing is also important: paddy should be planted in a well-soaked field with little standing water and then the depth of the water increased gradually up to six to twelve inches as the plant grows and flowers, after which it should be gradually drawn off until at harvest the field is dry. Further, the water should not be allowed to stagnate but, as much as possible, kept gently flowing, and periodic drainings are generally advisable for purposes of weeding and fertilizing. (Geertz, 1963; p. 31)

The labor and equipment needed to control the pulses of the wet rice system are enormous. Individual households are unable to provide enough labor during the peak periods—transplanting, uprooting, harvesting, and possibly plowing—and a complex nexus of interpersonal relations must be maintained to provide the labor (Hanks, 1972; pp. 50–51). For example, at harvest time

the grain must be cut, packed into sheaves, and carried to the threshing floor while the kernels still cling tightly to the inflorescence. If there is delay, the few days of dessication turn the connections brittle, so that grains drop in handling and are wasted. To finish in one day the job of cutting and moving the grain from the field to the threshing floor, one applies the formula: 1 man harvests 1/2 rai (about 1/4 acre, the amount of land an average water buffalo can plow in a day). Where neighborhoods exchange labor, returning one day of work for each day a neighbor spends in one's own field, a cultivator can count his days of labor by the number of persons he must invite to harvest his own field. (Hanks, 1972; p. 50)

One must suspect that, in addition to neighborhood groups, many social organizations exist to assist in finding the sufficient labor during the peak labor periods, particularly those organizations based upon kinship.

MANIPULATION OF ECOLOGICAL SUCCESSION

The ecologist Eugene Odum (1971) points out that it is to nature's advantage for an ecological system to increase *gross* production while it is to man's advantage to increase *net* production. One way of increasing net production is to manipulate the process of *ecological succession.* Succession is the process of change that takes place in an ecological community after colonization of a barren habitat. Early stages of succession (or *seres*) are dominated by *opportunistic* populations who can disperse rapidly into the newly opened habitat and cope with drastic changes in environmental conditions (Colinvaux, 1973; pp. 569–71). Opportunists allocate a large proportion of their available energy to *rapid growth* and *reproduction,* for example, annual plants. Both characteristics increase net production and are therefore useful to humans. After colonization of the habitat is well established, new habitats are not available and environmental conditions are less likely to undergo drastic change (Colinvaux, 1973; pp. 569–71). Opportunists are replaced by more efficient *equilibrium* populations with slow growth rates and minimal energy allocated to reproduction. As a result net productivity falls sharply, making later successional stages less useful to consumers such as humans.

One way to increase net production, then, is to keep ecological communities at early stages of succession. Fire is used for this purpose by several foraging groups (e.g., Stewart, 1956) and suggests that the strategy is ancient. In these cases fields and forests are periodically burned to encourage the growth of opportunistic populations that can be exploited. Farming, however, has been the most successful method of maintaining opportunistic food species. In this case mature communities are removed, by cutting, burning, and so forth, and are replaced by domesticated opportunists, especially annual plants. One widespread farming strategy is to take quick advantage of the opportunists and to get out before equilibrium species take over, called "shifting" farming or one of the 100 local names. Another important farming strategy, and one upon which modern civilizations are dependent, is to prevent succession from taking place by removing equilibrium species (e.g., by "weeding" or by using pesticides). One thing is common to all the strategies: Work must be done to bring about and maintain early stages of succession, whether by fire, humans, animals, or fossil fuels. Another problem that must be solved by all farming strategies is how to supply materials to the community. Only equilibrium communities have well-established material cycles; opportunists depend upon outside sources. Shifting farmers solve this problem by "pirating" the material reserves of equilibrium communities. That is, when a habitat is cleared for farming, the soil contains materials built up by an equilibrium community. The domesticated opportunists feed upon this supply until it is exhausted. The habitat is then allowed to undergo succession, reestablishing material cycles, so that it can be reused. Another solution is to add fertilizers at frequent intervals to provide a supply of materials to the opportunists.

Shifting Farming

Shifting cultivation is defined by

1. The clearing of forest openings with slash-and-burn techniques.

2. Planting cultigens using only the hoe or dibble stick and no tillage.

3. No use of fertilizers other than the wood ash deposited during slash-and-burn clearing.

4. No use of draft animals or fueled machines but complete dependence upon human labor.

5. No use of irrigation.

6. Fields are shifted frequently, and in a balanced system fields are rotated.

7. Fields are cultivated continuously for only a short period of time, usually no more than 2 or 3 years, and are then allowed to rest ("fallow") for a relatively long period, often 10 to 15 years.

The ratio between the period of continuous cultivation and the period of fallow is the most definitive characteristic of shifting farming. No other farming strategy is marked by fields that lie at rest longer than they are cultivated. The cultivation/fallow ratio varies from one ecological system to another because of physical and cultural factors. One important contrast is between *temperate forest* and *tropical forest* systems. Temperate forests have relatively long periods of cultivation, mostly because of nutrient cycling characteristics. Leaves and other litter are dropped on the ground and build up a thick humus layer because of slow chemical decomposition. Most of the nutrients in the temperate forest system are contained within the humus layer and not within the trees. When the trees are cut, the cultigens that replace them have a large nutrient reservoir to tap. By contrast, tropical forests have relatively short periods of cultivation because heavy rainfall and high temperatures greatly increase the decompositon rate of litter fall. In addition, nutrients are rapidly leached away because of the heavy rainfall. As a result, most of the nutrients in tropical forest systems are contained within the trees and not within the humus layer. The cultigens that replace the cut trees have a small nutrient reservoir and can be grown for only a short time.

The length of the cultivation period is also caused by the destructive method of shifting farming. Several studies have shown that shifting farmers can, and often do, have an adverse impact upon nature and that the strategy is not self-sustaining. The classic study of Iban farming in Borneo by J. D. Freeman (1955), for example, finds that keeping a field in production for two years in a row is sufficient to cause serious soil erosion and the conversion of forest to highly inflammable grassland. DeSchlippe's study of Zande farming (1956) gives data suggesting a similar transformation of natural woodland to grassland following overcultivation of fields. In general the techniques of shifting farming reduce crop yield with time because of (Netting, 1974, p. 26):

1. Multiplication of pests and diseases.

2. Increase of weeds.

3. Deterioration in the physical condition of the soil.

4. Erosion of the top soil.

5. Deterioration in the nutrient status of the soil.

6. Changes in the numbers and composition of the soil fauna and flora.

According to the zoologist Daniel H. Janzen (1973) the declining yields so often noted in the fields of shifting farmers are probably not due as much to the exhaustion of soil nutrients, the conventional explanation, as they are due to pest insects and competing weeds. This is particularly true in the tropics. Tropical plants produce toxic chemical compounds that combat insect pests; however, the tropical cultigens used by farmers have greatly reduced toxicity due to human modification. As a result tropical cultigens are highly susceptible to insect pests, and modern farming in the tropics could not exist without large quantities of pesticides.

Finally, there is more and more evidence to show that the ratio between the length of the period of continuous cultivation and the length of the fallow period is seldom based exclusively upon physical factors. Changes take place because of population pressure and cash-cropping demands, among other reasons. For example, Boserup (1965) proposes that population pressure causes a progressive shortening of the fallow period, and several studies support the contention. Also, data from the New Guinea highlands suggest that increasing population pressure is accompanied by a change from extensive shifting farming to more intensive shifting farming (Clarke, 1966). While Boserup's thesis has been sharply criticized for being oversimplified, it has not only opened communications between disciplines, but also has focused attention upon *time-depth* and *change* in studies of farming strategies.

Advanced Farming

Unlike shifting cultivation and other methods of simple farming, advanced farming "involves the almost total destruction of the preexisting natural ecosystem and its replacement by an artificial system with quite different structural properties and energy transfers" (D. Harris, 1973; pp. 393–94). Often called *agriculture,* such methods require a large energy expenditure from human labor and domesticated animals or a large input of fuels and materials to run the transformed or created system.

Labor-intensive agriculture (sometimes called "low-energy" agriculture) uses large amounts of human labor and does not depend upon fossil fuels. In pulsating systems (such as those in monsoon climates with sharp alternation between wet and dry seasons), large domestic draft animals are used to supplement human labor during critical planting or harvesting periods. The animals are supported by the net primary productivity of the system and do not represent outside energy subsidies. *Subsidized agriculture,* by contrast, does depend upon inputs of energy and materials from outside the system, inputs that drastically reduce the human labor requirement. At the same time subsidized agriculture is quite inefficient. The sociologist Fred Cottrell, in *Energy and Society* (1955) gives comparative data for wet-rice farmers in Japan and the state of Arkansas in the United States. Both farming systems yield about 50 bushels of rice per acre. However, Japanese wet-rice farmers do not use fueled machines while wet-rice farmers in Arkansas do. Energy expended by human labor alone is much greater in Japan: 90 man-days per acre as opposed to 14.1 man-days per acre in Arkansas. Cottrell converts these figures to horsepower-hours (using .1 horsepower per man per hour) and obtains the following values: Japan, 90 horsepower-hours; Arkansas, 1.41 horsepower-hours. It is apparent that the phrase "labor-intensive" is appropriate for the Japanese method of producing wet rice. Arkansas farmers, on the other hand, use a vast energy input from fossil fuels. The horsepower-hours of tractors, trucks, and electricity used to produce one acre of wet rice is calculated to be 805.43. The efficiency of wet-rice farming using fueled machines, then, is nearly nine times less than that of labor-intensive wet-rice farming but has an equivalent yield.

In *The Closing Circle* (1971), the ecologist Barry Commoner relates a personal experience about the cost of maintaining young successional stages through material subsidies. Corn yield in the state of Illinois (U.S.A.) was greatly increased during the period 1945–1965 by the application of inorganic nitrogen to the fields, as shown by the following data:

Date	Corn Yield (Bushels per Acre)	Fertilizer Nitrogen (Tons)
1945	50	Less than 10,000
1958	70	100,000
1965	95	400,000

An increment of only 100,000 tons of nitrogen was needed to raise average production by 20 bushels an acre, but the next yield increase of 25 bushels required 300,000 tons. It is obvious that efficiency of the corn fields has been considerably reduced. At the same time, Commoner notes that if local farmers receive yields of less than 80 bushels per acre, they cannot even meet expenses. As long as the cost of fertilizer is low, of course, inefficiency will mean very little.

What happens to inorganic nitrogen not used by the soil is of more immediate importance than inefficiency. The Illinois State Water survey collected data showing a great increase in the nitrate levels of rivers draining the Illinois farmlands during the period of time when fertilizer use had also increased greatly. In addition the water supply of Decatur was contaminated by nitrates. Commoner conducted experiments demonstrating that the nitrates must have come from fertilizer nitrogen rather than from another possible source, namely from soil, manure, or sewage sources. The competing sources all contain large quantities of nitrogen 15, a rare radioactive isotobe; fertilizer nitrogen does not. Rather, it contains nitrogen 14, the same nitrogen isotobe that makes up 99.6 percent of all natural nitrogen. The contaminated water measured by Commoner and his colleagues was low in nitrogen 15, thus suggesting that the source of the new nitrates was fertilizer nitrogen.

How do nitrates in such large quantities affect man? Commoner reports that studies at the University of Illinois show increased mortality of girl infants when nitrate levels of water supplies were high. (The death rate was 5.5 per 1000 when nitrate levels were high and only 2.5 per 1000 when nitrate levels were low.) Why? According to Commoner,

Nitrate itself appears to be relatively innocuous in the human body. However, it can be converted to nitrite *by the action of certain intestinal bacteria, which are often more active in infants than adults. And nitrite is poisonous, for it combines with hemoglobin in the blood, converting it to methemoglobin and preventing the transport of oxygen by the blood. An infant thus affected turns blue and is in serious danger of asphyxiation and death.* (1971, p. 82)

REMOVAL OF LIMITING FACTORS

Energy flow through some ecological systems is severely inhibited by a shortage or an overabundance of materials or by intolerable physical conditions. Humans have often used cultural methods to alleviate the stress and increase flow through the system. A good example of this kind of modification is the temperature control used in greenhouses. However, the energy subsidy needed to control temperature is so large that greenhouses over large areas are presently impractical. On the other hand, the energy subsidy required to remove *water* limitations from large areas is not as large, and water control is not only a widespread but is also an old practice. Water stress results both from too little and too much water so that cultural factors have been directed toward both kinds of problems.

Too Little Water

Numerous methods of adding water to increase gross production have been devised by humans. Bennett (1974) lists irrigation systems, storage ponds and reservoirs, rainfall catchment basins, terracing of slopes to hold water to increase aquifer recharge, pot irrigation, water table excavation, aqueducts and tunnels, and wheels, levers, and buckets to lift water. Of these, *irrigation systems* are the most familiar to the reader and include any devices that divert water to improve plant growth. The most simple irrigation system from the standpoint of technology is a ditch dug through a natural levee of small streams so that water can escape. It has been suggested that such a system may have been the earliest form of irrigation in Mesopotamia, established by at least 7500 years ago (Flannery, 1969). A slightly more elaborate system was used by the Eastern Mono of California to increase the standing crop of wild seeds and tubers. Julian Steward (1930) described one Owens Valley system in some detail. Owens River, flowing along the valley floor, is intersected at the northern end of the valley by Bishop Creek, a large creek flowing out of the Sierras. On each side of Bishop Creek were large fields of wild seeds and tubers, the largest of which was four miles in length and 1½ miles in width. An irrigation system was constructed to divert water from Bishop Creek into the two fields. The system included a single dam one mile below the point where Bishop Creek emerged from the Sierras and two ditches running from the dam, one to each of the fields. Only one plot was watered each year, during the spring, and the dam was destroyed each fall to allow the water to return to its original course. In the spring about 25 men constructed a new dam of stones, wood, and mud, and an "irrigator" was chosen. The irrigator was responsible for directing the flow from the main diversion ditch so that the entire field was watered during the irrigation period. This involved no more than building small "check dams" and subsidiary ditches to change the direction of flow.

A similar irrigation system was described for the Papago of Southern Arizona by Edward Castetter and Willis Bell (1942). Fields of domesticated plants were situated on the broad alluvial fans at the mouths of Arroyos. Rainfall runoff in the mountains flowed down the arroyo canyons and spread out in a broad sheet upon reaching the alluvial fan. Papago irrigation systems were designed to collect the runoff and divert it to the fields. Shallow ditches, low embankments, and brush dams were constructed as part of the system and, on occasion, a fairly large ditch would be constructed from the arroyo directly to the fields. No attempt was made to build reservoirs to hold water (except historically), and irrigation took place only when arroyos ran with water. In contrast to the Owens Valley Mono, each Papago family constructed and maintained its own ditches and dams, unless a ditch supplied water to fields belonging to more than one family.

Larger irrigation systems often required no more advanced technology than the simple Mono and Papago systems but demanded a large labor force to build and maintain the system. The labor requirements of maintenance alone so impressed the historian Karl Wittfogel (1957) that he theorized the origin of some states in the managerial needs of large-scale irrigation networks. Unfortunately, accumulating data in at least Mesopotamia and Mesoamerica suggest that early states evolved without any role in irrigation management. Early irrigation systems were small, based upon simple technology, and maintained by families. The Mono and Papago systems are good examples of how this might have been done, along with cutting of natural levees to extend the growing season. On the other hand, there can be no doubt that state management was necessary for more complex irrigation systems. The Inca used impressive engineering skills to cut irrigation canals from solid rock, to build masonry aqueducts across gorges and high terraces on mountain slopes, and to build storage reservoirs. Control of *mita* or *corvee* labor was necessary for the construction and maintenance of these devices. Furthermore, on the Peruvian coast fortifications were constructed at the narrow entrances to valleys in order to control access to rivers flowing into the valley. Although perhaps constructed prior to the Inca, the size of some irrigation systems on the coast is quite impressive indeed. The La

Cumbre canal in the Chicama Valley on the North Coast extends about 75 miles, from the headwaters of the Chicama River to its mouth. Of course none of these systems would match the proposed irrigation canals linking the Pacific Northwest of the United States with the arid Southwest!

Water-Table Excavation is used in dry habitats with relatively high water tables or in less dry habitats where exceptionally wet field conditions are desirable. The method consists of excavating a pit down to the water table and planting crops at the bottom of the pit. On the south and central coast of Peru dry valleys without rivers are cultivated in this manner (Parsons 1968). The methods used is called *mahamaes* or *pukio* cultivation. Mahamaes gardens are grown on a strip near the coast where the water table is highest and closest to the surface. (The strip was about three kilometers wide in the Chilca Valley near Lima.) Each garden is excavated to a depth of two to five meters. Natural water at this depth provides plenty of moisture to grow maize, beans, squash, and a number of other crops. An individual garden is not large, ranging in size from 20 meters to 100 meters square.

Jacques Barrau (1965) describes a similar technique used by the atoll dwellers of the Pacific Islands. Atolls contain groundwater that is almost entirely salty or brackish except for a few lenses of fresh water near the center of the islands. Pits are excavated into the center of the atoll down to the fresh water lens. Bottomless baskets, woven from twigs, are placed in the mud at the bottom of the pit and are filled with miscellaneous organic material. A form of taro *(Cyrotosperma chamissonis)* is then planted in the baskets. Barrau observes that the floating baskets almost duplicate the swampy lowland environment from which the plant originated.

High water tables are also taken advantage of in a watering technique known as *riego a brazo* or "pot-irrigation". This practice has been documented among the Zapotec in highland Mexico (Flannery et al., 1967) and consists of digging shallow wells in the midst of cultivated fields. A three gallon pot is used to draw water from each well, and the water is poured around individual plants. Pot-irrigation makes it possible to grow crops as long as the temperature is not limiting, and up to three crops a year are grown in the Valley of Oaxaca using this method. Of course, the method depends upon a high water table and can be applied only in a few areas. A similar technique for irrigating has been described in the American Southwest, but in this case pots are used to draw water from a nearby river.

No discussion of watering systems is complete without at least mentioning the famous *qanat* or *foggara* systems of the Middle East. Long underground conduits are used to transport water from high water tables at the base of mountains to fields of cultivated crops. Shallow wells tap into the conduit at frequent intervals, thus giving a usable irrigation ditch with minimum evaporation.

Planning the location of fields is often the only technology required to obtain water in dry habitats. In western Iran early dry-farming communities were situated near the margins of swampy areas, thus taking advantage of a naturally high water table. Flood plains were important determinants of where dry-farming communities would be situated in many parts of the world. Thus, in the southern deserts of North America, the Yuman tribes of the Lower Colorado and the Cahita tribes occupying several rivers in Sinola relied on annual flooding of river valleys for agriculture. The Cahita actually grew two crops a year by using summer and winter floods. In Mesopotamia early dry-farming settlements in the Deh Luran valley were situated upon the low rises surrounding the central depression of the floodplain. It seems almost certain that floodplain agriculture was being practiced.

Too Much Water

Increasing intensity of land use often involves bringing previously inaccessible land into production, as well as growing more crops on the same land. Overcoming the obstacle of too much water is a common way of bringing this about, and swamps are the areas generally recovered.

Ridged fields duplicate natural levees. On many of the floodplains of major rivers in the Amazon Basin, natural levees take the form of parallel ridges that also run parallel to the river banks. Tropical forest cultivators farmed the tops of these ridges. Artificially constructed ridges similar to natural levees were used in several parts of tropical America to make farming possible in low swampy areas. Ridged fields were particularly common in savannah regions subject to annual flooding, including the Guayas Basin on the southern coast of Ecuador, the Mompos Depression in northern Columbia, the western coast of Surinam, and the Llanos de Mojos of northeastern Bolivia. In addition ridged fields have been documented on the western shores of Lake Titicaca in the Andean highlands and in the Usumacintla River Basin in southern Mexico. The most elaborate system is situated in the Llanos de Mojos, and the archaeologist Donald Lathrap (1970) has argued that all ridged field systems in South America originated here. The Llanos de Mojos is covered with a minimum of 10,000 raised fields distributed over 15,000 acres and probably contains several hundred thousand fields covering 50,000 acres. The fields are of several sizes, the largest being 300 meters long and 27 meters wide. During flooding, the fields are above water but the space between the fields begins to silt-in. To avoid complete filling, the spaces are cleaned periodically, and the silt is dumped on top of the ridges. Construction of ridged fields appears to have been a partial solution to increasing population pressure in lowland Mexico and Guatemala. Dennis and Olga Puleston (1971) have argued that the preclassic Maya farmed natural levees after their initial movement into the Usumacinta and Candelaria River Valleys but later added ridged field systems. Such fields increase the amount of arable land on the floodplain, and their construction probably represents a solution to increased population pressure.

Chinampas are small islands artificially constructed in shallow bodies of water. The Aztec of the Valley of Mexico relied extensively upon this method, from whom comes the name. Thick mats of water plants (mostly reeds and water lilies) were cut from lake shores and piled together. Mud from the lake bottom was spread over the top and willow trees were planted to fasten the new island to the lake bottom. Thus, while today they are commonly called "floating gardens," they do not float. *Chinampas* were among the most productive fields in the New World and were maintained by adding new soil from the lake bottom. Jacques Barrau (1965) describes a system used on many high islands of Polynesia that is quite similar to the Mexican *chinampas*. Mud ridges or mounds are constructed in shallow ponds, and taro is planted on top.

ORGANISM MODIFICATION

Increasing net production has limited advantages if only a small part of the food plant or animal can be eaten. In many cases an uneatable part can be shunted to a "middle-man" capable of eating that part and converting it into food usable by man. For example, domestic cattle can be fed wheat straw or corn fodder, parts that man cannot eat, and the cattle in turn can be used for milk, butter, and cheese, "tapped" for blood, or butchered for flesh. A more direct method is manipulating the genetic characteristics of the plant or animal to increase the percentage of usable parts. Kent Flannery (1965), using the work of the Danish botanist Hans Helbaek, suggests several ways in which early domesticated plants in the Middle East may have been manipulated. The wild progenitors of wheat and barley shatter their seeds readily upon contact with the plant to facilitate reproduction; the seeds thus dispersed have very tough husks for protection until germination can take place during an appropriate season. The consequences for man are apparent. Harvesting with sickles or other tools would cause the seeds of individual plants to scatter over the ground, making collection difficult. When those seeds that could be collected were prepared for eating, the tough husk rendered conventional thrashing techniques almost useless. Intense artificial selection for less brittle dispersal mechanisms and husks that could more easily be removed undoubtedly accompanied early attempts at domestication of these cereals. Of course, as Flannery observes, another option is to roast the grain prior to husking, thus making the husk easy to remove and providing an excellent example of how cultural behavior can bypass the necessity for biological change. In this case both genetic change and cultural methods were used to solve this problem. Another difficulty with the wild progenitors of barley and wheat is their adaptation to hillsides and slopes although farming is best in flat river valleys. Moreover, local varieties of wild wheats and barleys were narrowly adapted to particular climates, soils, and elevations, making it exceedingly difficult for early farmers to expand into a variety of ecological zones. Genetic change through hybridization improves the tolerance of local strains for exotic ecological zones and makes it possible for wheat and barley farming to expand, not only into the valleys of rivers and streams but also into new climates, soils, and elevational zones. A similar change has been suggested for maize in the New World. Hybridization and back-crossing with wild progenitors were also responsible for increasing the overall yield of wheat and barley seeds and the maize ear. In fact, modern agricultural geneticists continue to manipulate the genetic structure of crops to improve their usefulness to man. One result of this manipulation has been to make the domestic crop completely dependent upon man for reproduction and for proper growing conditions.

What about animals? Stock improvement in recent years to produce fast-growing or heavy cattle for market purposes suggests the importance of artificial selection for parts usable by man. This history of at least one domestic animal—the sheep—is marked by artificial selection (Flannery, 1965). The wild progenitors of sheep were hairy rather than woolly and not at all suited for wool production. A thin layer of wool insulated wild sheep under a thick coat of hair. Early domestic sheep still retained this characteristic, but through time artistic representations indicated that the amount of hair gradually declined and the amount of wool increased, finally resulting in the predominantly woolly sheep with which we are all familiar. The hair-wool relationship has been reversed, and modern domestic sheep have only a thin layer of hair underneath a thick layer of wool. Similar genetic change under intense artificial selection was undoubtedly responsible for the present importance of domestic animals to man.

SUMMARY

Humans *intentionally* change ecological systems in which they participate. They do so by using cultural devices to "capture" pulses of energy and materials, to take advantage of the process of ecological succession, to remove physical stresses (such as water shortage), and to modify resources. Few would question the significance of these methods to the way that humans solve problems of energy and materials acquisition. But in addition, by using culture to remove or reduce the impact of predators, disease, and meteorological or geological disasters, humans ameliorate the problem of hazards. Civilization has been the culmination of human efforts to manipulate and create ecological systems. In the words of Colin Renfrew,

> *we can see the process of the growth of a civilisation as the gradual creation by man of a larger and more complex environment, not only in the natural field through increasing exploitation of a wider range of resources of the ecosystem, but also in the social and spiritual fields . . . Civilised man lives in an environment very much of his own creation . . . All the artefacts which he uses serve as intermediaries between himself and this natural environment and, in creating civilisation, he spins . . . a web of culture so complex and so dense that most of his activities now relate to this artificial environment rather than directly to the fundamentally natural one. (1972, p. 11)*

It is becoming increasingly clear that man's role as a creative force in ecological systems has introduced a wide range of *new* problems and unexpected side-effects. Salinization and schistosomiasis from irrigation schemes, social and psychological pathologies from cities and economic development schemes, the buildup of toxic chemicals in the air, soil, and water from artificial fertilizers and pesticides—all of these are problems created by man during the quest for solutions to yet other problems. Human evolutionary success will ultimately depend upon how well human-created problems can be coped with.

SIX THE HUMAN ECOLOGICAL NICHE

The relationship of a human population to other organisms in an ecological system can be systematically studied with the concept of the ecological *niche.* Although in use for a long time, the niche concept received its modern definition from Charles Elton (1927). He viewed it as a "role" in a food web, a distinctive feeding strategy that sets one organism apart from another. In terms of energetics, the niche is an organism's share of the limited energy and nutrients available in an ecological system. For humans or any other consumer organism that must "feed" for a living, the available energy and nutrients can be viewed as a *variety* of food resources in the form of biomass or materials. The resources vary in size, color, distribution in space and time, temperature, mobility, and so forth. Those variants upon which an organism feeds is its niche.

THE NICHE HYPERVOLUME

G. E. Hutchinson (1965) suggested that it is useful to characterize the niche as a *volume* in a multidimensional Euclidean hyperspace. Theoretically, the hyperspace is made up of a large number of resource variables, each of which is defined as a *dimension* of the hyperspace. In practice the hyperspace has only a few dimensions, each of which is a "limiting" resource variable that is useful for separating one niche from another (Levins, 1968; Roughgarden, 1972). Figure 6.1 illustrates a hyperspace with only three dimensions: resource size, resource color, and resource shape. If we assume that each dimension varies infinitely, a Euclidean *hypervolume* is defined with the shape of a *cube.* However, an organism

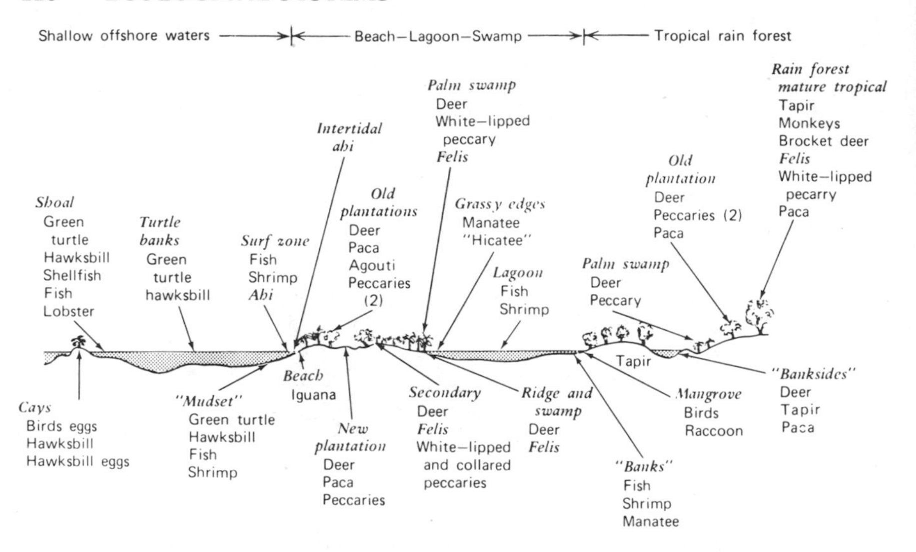

FIGURE 6.1 Miskito microenvironments. (From Bernard Nietschmann, *Between Land and Water*, New York: Seminar Press, 1973. Copyright © 1973 by Academic Press, Inc. Reprinted by permission of the author and publisher.)

makes use of only a portion of this cube by exploiting only resources within a definite range of size, color, and shape. The segments representing the ranges of the three resource dimensions actually utilized by an organism are the three sides of a cube that represents the organism's niche or life style. Simple calculation of the volume of the cube provides, therefore, a single index for three variables of the ecological niche. The addition of a fourth, fifth, or sixth dimension to the hyperspace makes the calculation somewhat more difficult, but the procedures are straightforward.

Hutchinson also points out that the niche of an organism can be viewed in at least two different ways: as a theoretical way of life in which no *competition* exists and as an actual way of life under conditions of competition. The first he calls the *fundamental* niche, and the second the *realized* niche. Consider the following example. Suppose that a human group occupied an isolated valley in which there were no other humans. Food resources in the valley were both plentiful and varied, and the group ranged widely, using all of the resources equally. However, with the passage of time, the group grew in size until it was impossible for everyone to use all of the resources at the same time. An arrangement was worked out to prevent squabbling. The group divided into a few smaller subgroups, each of which used only *some* of the available resources. This solution solved the problem of population pressure until one day a *new* group moved into the valley, a group with food habits similar to the initial group. Now the problem was intensified. Not only was competition for resources caused by normal population growth but also by other groups. The problem was finally relieved by spatial segregation, with the indigenous groups using only the southern half of the valley, and the new group using the resources in the northern half of the valley. In this example the fundamental niche and realized niche of the ancestral group were identical. With the addition of population pressure, however, the realized niche of each subgroup was only a fraction of its fundamental niche, caused by the differential use of resources. Finally, the addition of competition from other groups forces an even further narrowing of the realized niche by spatial exclusion.

MICROENVIRONMENTS

How are human niches to be studied in practice? A useful approach is to equate the niche with the *spatial* use of subsistence resources. Human populations divide their habitat into distinct resource clusters called

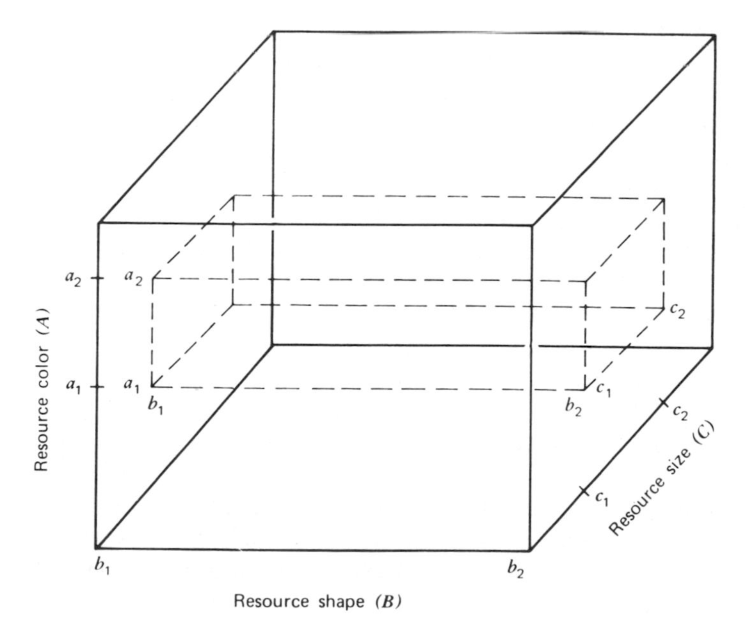

FIGURE 6.2 The niche hyperspace. The total resource space available is *ABC*. The realized niche space is $(a^2 - a^1)$ times $(b^2 - b^1)$ times $(c^2 - c^1)$.

microenvironments (Coe and Flannery, 1964). A microenvironment is very similar to the *biotope* as used in biology, defined as a "microhabitat or the smallest natural area or space characterized by a particular environment" (Cain, 1966, p. 47).

The geographer Bernard Nietschmann (1973) gives an example of microenvironments as defined by the Miskito Indians of the eastern coast of Nicaragua. The ecological framework of the Miskito consists of four biomes: tropical rain forest, pine savannah, beach-lagoon-swamp, and shallow offshore waters. However, the biomes are perceived as diverse and are exploited unevenly. The pattern of utilization can best be described as a "hunting and fishing focus," joining biotopes containing concentrations of terrestrial game animals, such as deer, peccaries, and iguana, and freshwater and marine species of fish, shrimp, and green turtles. Figure 6.2 illustrates the major microenvironments of the Miskito. The *number* of distinct sets of resources is large, and the reader may be curious about the impact of a large number of microenvironments upon the ecological niche of the Miskito. Other things being equal, if the number of microenvironments used for subsistence is large, the number of resource varieties is large and the niche is said to be *generalized*. By contrast dependence upon a few microenvironments implies a *specialized* niche, because only a few resource types are used.

However, "other things" are seldom equal. All microenvironments do not make the same contributions to subsistence. Nietschmann spent a considerable amount of time collecting data on the relative importance of Miskito microenvironments, and some of the results are given in Table 6.1. Inspection of the table will show that a small number of marine and estuary microenvironments contributes a much larger *quantity* of meat, 87 percent of the total meat diet, than does the larger number of terrestrial microenvironments. From this point of view the Miskito have a much more specialized niche than would be suggested by their complex network of microenvironments.

Quantity is only one measure of the importance of microenvironments. Resource *quality* is a measure of how critical each resource is to a human population and must be considered in evaluating the importance of microenvironments. For example, a microenvironment contributing a large quantity of starchy roots may not be nearly as important as a microenvironment contributing a small quantity of high-quality protein. "Timeliness" is another measure of resource quality, as suggested by the geographer Homer Aschmann in describing the subsistence pattern of the Indians of the Central Desert of Baja California.

The seasonal availability of a particular food was probably of more significance than the amount present. The carrying capacity of the area, in terms of a human

TABLE 6.1 RELATIVE CONTRIBUTIONS OF MICROENVIRONMENTS TO MISKITO SUBSISTENCE

	White-Lipped Peccary	White-Tailed Deer	Green Turtle	Hawksbill	Iguana	Fish	Shrimp	Hicatee	Miscellaneous (16)	Total
Ocean										
Shoals ("sleeping grounds")			27,150	1100		300			Whelks, Conch Lobster [a]	28,550
Banks (feeding grounds)			43,205	855						44,060
"Mudset"			6,250	50		450	400			7150
Surf zone									*Ahi* 700	700
Intertidal									Birds, bird eggs [a]	
Cays										
Land										
Beach		120			201					321
Old plantations	800	820								1620
New plantations	150	1600							Collared peccary 40	1790
Ridge and swamp	50	250								300
Marsh		300								300
Palm swamp	3050	1080								4130
Scrub thicket	100									100
Secondary	2000	600							Collared peccary 40	2640
Mature rain forest	300								Brocket deer 40	340
Gallery forest	180	360			56					596
Bankside ("vega")		520			44				Tapir 200	764

TABLE 6.1 *(continued)*

	White-Lipped Peccary	White-Tailed Deer	Green Turtle	Hawksbill	Iguana	Fish	Shrimp	Hicatee	Miscellaneous (16)	Total
Freshwater										
Lagoon						400				400
Creek mouths						300	400			700
Shallow borders						600	350	800		1750
"Banks"						700	1850		Manatee 200	2750
Sandbars								105	Caiman, iguana, Crocodile, hicatee eggs [a]	105
Creeks						200		95		295
Grass flats								1355		1355
Totals	6630	5650	76,605	2005	301	4100	3000	2355	1220	101,866

Source. Bernard Nietschmann, *Between Land and Water,* New York: Seminar Press, 1973. Copyright © 1973 by Academic Press, Inc. Reprinted by permission of the author and publisher.

[a] *Not measured.*

population which made little effort to store food, must be stated in terms of what was available in the poorest season out of several years, not in terms of average food supply. Consequently, a food available only in small quantity and ordinarily ignored may be the one that at critical moments prevented starvation. A consideration of only the ten or twenty most important foods may miss this critical aspect of the food economy. (1959, p. 78)

Thus, a stable microenvironment, one that does not fluctuate greatly from year to year, may be more important to a feeding strategy than an unstable microenvironment.

QUANTIFICATION OF THE HUMAN NICHE

The niche of a human group can be quantitatively measured by calculating the amount of resource *variety* that is used for subsistence. As used here, subsistence variety is "richness," the number of different resources used, and "evenness," the degree to which each is depended upon. Conventionally, the amount of variety is called "niche width" or "niche breadth" with the stipulation that a niche with a narrow wide, or low variety, is specialized, while a niche with a broad width, or high variety, is generalized. Human niche width can be calculated from the following formula:

$$\text{niche width} = \frac{1}{\sum_{i}^{n} (pi)^2}$$

where pi is the proportion of the total subsistence contributed by resource i and n is the total number of resources used for subsistence (modified from Levins, 1968, p. 43). Table 6.2 shows how niche width is calculated from "pooled" subsistence data given by Rogers for the Mistassini Cree of Eastern Canada without regard to space or time. Using similar subsistence data, I have calculated niche widths for several other human populations. The values are given in Table 6.3.

In addition to *total* variety in the resources used for subsistence, niche width can also be calculated from *spatial* variety alone. Spatial variety is measured by the number of microenvironments exploited for subsistence and the degree to which each microenvironment is depended upon. If the number is large and dependence is equal, the niche is generalized and its width is wide. On the other hand, if the number of microenvironments is small and they are not used equally, the niche is specialized and its width

TABLE 6.2 NICHE WIDTH CALCULATIONS FROM TOTAL RESOURCE VARIETY FOR MISTASSINI CREE, CANADA[a]

Resource	Biomass (lb)	*pi*	*(pi)*²
Moose	4000	0.448	0.202
Caribou	1500	0.168	0.029
Bear	210	0.024	0.000
Beaver	2120	0.237	0.058
Hare	114	0.013	0.000
Muskrat	240	0.027	0.001
Porcupine	60	0.007	0.000
Mink	33	0.004	0.000
Squirrel	8	0.000	0.000
Marten	5	0.000	0.000
Otter	110	0.012	0.000
Loon	44	0.005	0.000
Geese	67	0.008	0.000
Ducks	231	0.026	0.001
Ptarmigan	150	0.017	0.000
Spruce grouse	38	0.004	0.000
Ruffled grouse	1	0.000	0.000
Owl	1	0.000	0.000
		$\sum_{i}^{n} (pi)^2$ –	0.291
	Niche width –	1/0.291 –	3.436

Source. Donald L. Hardesty, "The Niche Concept: Some Suggestions for Its Use in Human Ecology", *Human Ecology*, 3, 1975.

[a] *Data from Rogers, 1972 104, Table 3-1.*

is narrow. Table 6.4 illustrates the method of calculating niche width from spatial variety, using data given by Nietschmann for the Misquito of eastern Nicaragua.

Finally, niche width can be calculated using different resource variables. For example, Table 6.5 gives the niche width of three human populations calculated for two resource variables: caloric content and protein content. The two width measurements are quite different in all three cases. Both the Tsembaga Maring and Tasbapauni Miskito are much more specialized for sources of protein than for calories. On the other hand, the Bayano Cuna are much more specialized for calories than for protein. If the reader will recall, Levins (1968), MacArthur (1960), and other ecologists have stressed the importance of *critical* resources in the partitioning of a resource hypervolume into niches. Thus, the significant

TABLE 6.3 NICHE WIDTHS FOR SEVERAL HUMAN SOCIETIES

	Niche Width
Food-collecting societies	
Kostenki IV-2 Pleistocene hunters (caloric data from Klein, 1969, p. 222, Table 38).	1.569
Mistassini Cree, Canada (biomass data from Rogers, 1972, p. 104, Table 3-1).	3.436
Central Desert Indians, Baja, California (biomass data from Aschmann, 1959, p. 103).	7.874
Food-producing societies	
Chimbu, New Guinea (biomass data from Rappaport, 1968, p. 73, Table 8).	1.685
Kapauku Papuans, New Guinea (biomass data from Pospisil, 1963, p. 444, Table 24).	1.698
Busama, New Guinea (biomass data from Rappaport, 1968, p. 73, Table 8).	2.206
Kavataria, New Guinea (biomass data from Rappaport, 1968, 73, Table 8).	2.892
North Kavirondo Bantu, Maragoli, Africa (biomass data from Wagner, 1954, p. 20).	4.651
Tasbapauni Miskito, Nicaragua (caloric data from Nietschmann, 1973, p. 220, Table 32).	5.283

Source. Donald L. Hardesty, "The Niche Concept: Some Suggestions for Its Use in Human Ecology," *Human Ecology*, 3, 1975.

measurement of niche width in the Tsembaga Maring and Tasbapauni Miskito populations depends upon whether calories or protein resources are limiting. Of course, another method of calculation is to treat both calories and protein as dimensions of a resource hypervolume, following Hutchinson (1965), and to calculate a single measurement of niche hypervolume. However, the use of a single critical resource is probably better suited to available human ecological data.

INTERPRETATION OF NICHE WIDTH

Although the preceding section compares the niche widths of several human populations, they are not always comparable. The width of a niche is a relative measure of an organism's share of available resources. It is comparable to other niches *only* when they are part of the same ecological system. Consequently, the niche widths of human populations par-

TABLE 6.4 NICHE WIDTH CALCULATION FROM SPATIAL RESOURCE VARIETY FOR TASBAPAUNI MISKITO[a]

Biotope	Meat Biomass (lb)	*pi*	$(pi)^2$
Ocean			
Shoals	28,550	0.280	0.078
Banks	44,060	0.433	0.188
"Mudset"	7,150	0.072	0.005
Surf zone	1,150	0.011	0.000
Intertidal	700	0.007	0.000
Land			
Beach	321	0.003	0.000
Old planatations	1,620	0.016	0.000
New plantations	1,790	0.018	0.000
Ridge and swamp	300	0.003	0.000
Marsh	300	0.003	0.000
Palm swamp	4,130	0.041	0.002
Scrub thicket	100	0.001	0.000
Secondary	2,640	0.026	0.001
Mature rainforest	340	0.003	0.000
Gallery forest	596	0.006	0.000
Bankside	764	0.008	0.000
Freshwater			
Lagoon	400	0.004	0.000
Creek mouths	700	0.007	0.000
Shallow borders	1,750	0.017	0.000
"Banks"	2,750	0.027	0.001
Sand bars	105	0.001	0.000
Creeks	295	0.003	0.000
Grass flats	1,355	0.013	0.000
		$\sum_{i}^{n} (pi)^2$ =	0.275
	Niche width =	1/0.275 =	3.617

Source. Donald L. Hardesty, "The Niche Concept: Some Suggestions for Its Use in Human Ecology," *Human Ecology*, 3, 1975.

[a] *Data from Nietschmann 1973: 169, Table 22.*

ticipating in different ecological systems are not comparable. The usefulness of the niche width measurement is in understanding the energetic relationships between a human population and its biotic environment (and this may include other human populations). Precise evolutionary studies of human ecology definitely requires this type of quantification. What are some of the problems that might be studied? An important one

TABLE 6.5 DIFFERENCES IN WIDTH BETWEEN TWO NICHE DIMENSIONS

Society	Calories	Protein
Tsembaga Maring, New Guinea (data from Rappaport, 1968, pp. 280–281, Table 26).	5.917	0.991
Tasbapauni Miskito, Nicaragua (data from Nietschmann, 1973, p. 220, Table 32).	5.283	2.688
Bayano Cuna, Panama (data from Nietschmann, 1973, p. 220, Table 32).	1.586	3.423

Source. Donald L. Hardesty, "The Niche Concept: Some Suggestions for Its Use in Human Ecology," *Human Ecology*, 3, 1975.

is the *relative abundance* of humans and other organisms in an ecological system and the impact of evolutionary change upon it. For example, MacArthur (1960) used niche width differences to explain the relative abundance of equilibrium species (see Cohen 1968 for some alternative explanations). He was able to show mathematically and to support with empirical data that the characteristic log-normal distribution of abundance can be produced if:

1. A critical resource is responsible for different niches.

2. The niches conform to the principle of competitive exclusion (Hardin, 1960) and do not overlap.

3. The width of the niches is caused by random partitioning of the critical resource.

Since wide niches represent a large share of a critical resource, occupant species are abundant. By the same reasoning those species with a narrow niche are rare. From this point of view, a useful study of the relative abundance of human populations and other meat eaters in tropical rain forests could be made. Protein is a critical resource in such ecological systems (Gross, 1975), so that the partitioning of available protein should be closely correlated with abundance. Furthermore, if a population expands its niche, its resource share increases and so does its abundance. There is no question that human populations have used "interference" methods to increase its niche width. Perhaps the best example is monocrop farming in which the energy flow of an ecological system is channelled into one or a very few types of resources important to human feeding and competing organisms are eliminated by pesticides, herbicides, weeding, and so forth.

SUMMARY

A human ecological niche is a feeding strategy of a human group. It defines a set of resources within an ecological system that is essential to the group's survival and is measured by subsistence variety. The niche of a human group is not static but changes with the process of adaptation. According to Richard Wilkinson, adaptive change occurs either to improve the "fit" between a human group and a constant environment or to cope with new problems caused by a change in the environment (1973). The human ecological niche changes in the same way, a change that is brought about by social and cultural means. Technological innovation, a new political alliance, a new labor union—all are ways of changing the relationship of a human group to its physical and social environment and, therefore, of changing the niche.

PART TWO HUMAN POPULATION ECOLOGY

SEVEN
POPULATIONS

Part 1 has dealt with humans as part of an ecological system. Because of its emphasis upon the broad interrelationships among organisms, this approach is called *synecology*. Another approach, *autecology*, is more concerned with the interactions that explain the abundance, distribution, and composition of specific populations. Part 2 looks at humans from the autecological point of view, focusing upon the human population. The population concept has many advantages for human ecological studies. First, it is a more or less bounded unit that can be quantified (Vayda and Rappaport, 1968, p. 494). Second, its characteristics are often correlated with features of the physical, biological, and social or cultural environment and change during the process of adaptation. Consequently, human adaptation can be studied in part by means of the population. Figure 7.1 illustrates some of the adaptive interactions. Third, the population, not the individual or the ecological system, is the principal unit of evolutionary change. Finally, the population is important to synecological studies using the niche concept because it is the population that occupies the niche (Colwell and Futuyama, 1971, p. 575).

POPULATION CONCEPTS

Prior to describing its interactions, a human population must be identified. First of all, an appropriate *concept* of population is defined. Ecologists most often use the *breeding population*, ranging from the species for an organism with a restricted geographical range to the deme (local breeding population) for widespread species. Since the breeding population defines the gene pool, it is most appropriate for organisms that depend upon genetic mechanisms for adaptation. Vayda and Rappaport (1968) have proposed that a similar concept of population be used for human ecological studies. Since the local breeding population, the deme, is often equivalent to a human settlement, for example, a camp, village, or

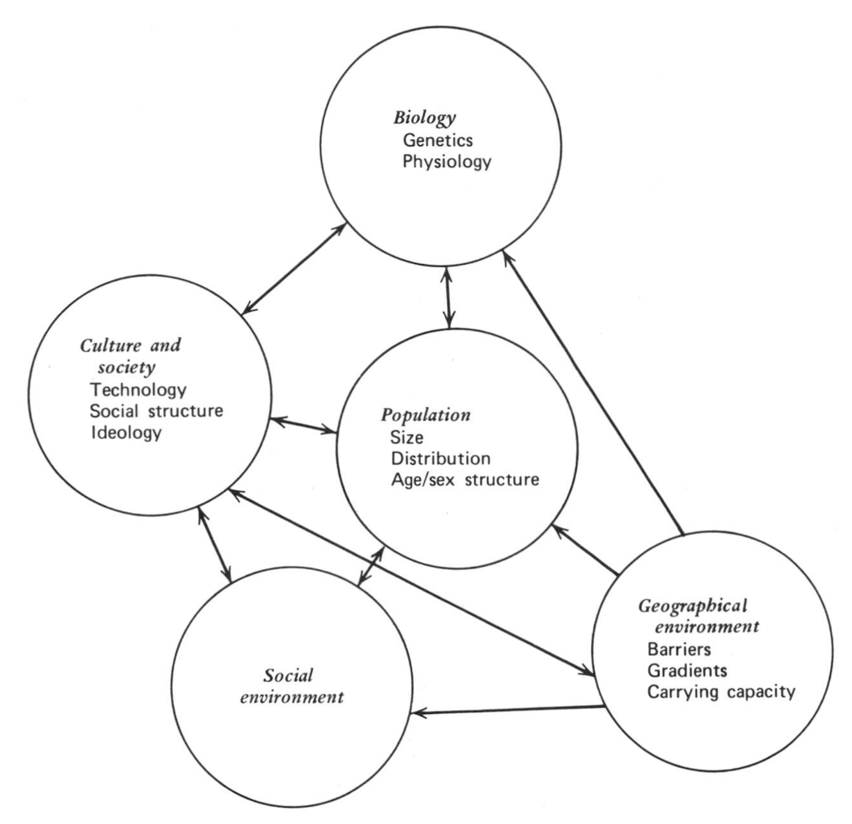

FIGURE 7.1 Some ecological relationships of human populations.

town, it can be easily identified and described. Furthermore, the local community is often integrated by kinship, social, economic, and political ties into an effective adaptive organization. However, the deme is not always the appropriate unit. The effective "ecological" population is often a vast network of local groups integrated by economic, social and political ties (Hawley, 1973). The modern nation–state is an example, as is the neighboring bands of foragers who cooperate to reduce the economic impact of an unexpected local food failure. For human ecological studies, then, it is best to recognize a "hierarchy" of populations, from a deme to a geographically large and diverse organization, any one of which may be appropriate for a particular problem but all of which are part of the adaptive process.

POPULATION BOUNDARIES

After an appropriate concept of population has been selected, its *geographical boundaries* must be delimited. Boundaries are easy to identify if the population is geographically isolated and separated from other populations by barriers. Thus, the human populations on small islands present no problems. Neither are there problems if the population concept being used is synonomous with a nation–state or similar unit with well-defined territorial boundaries. Unfortunately, many populations do not have distinct geographical boundaries. Individuals or settlements are continuously distributed in space, broken only erratically by features of the landscape. What are the boundaries of these populations? If the concept of the breeding population is appropriate, boundaries can be defined by geographical patterns of marriage. The geographical region within which marriage partners are found and reproduction takes place defines the gene pool boundaries as well.

Marriage patterns have a geographical distribution determined by (1) *marital movement,* any geographical movement involved in the selection of mates; (2) *postmarital movement,* the geographical movement determining where reproduction takes place after marriage; and (3) *parent/offspring movement,* the total movement involved in the selection of mates and reproduction (the combined marital and postmarital movement). Marital and postmarital movement can be measured by several indices developed by anthropological demographers. *Marital distance,* the distance between the birthplaces of the marriage partners, is a useful measure of marital movement. For a population marital distance is calculated by averaging the distances between the birthplaces of all marriage partners (Harrison and Boyce, 1972, p. 132). A more precise measure of marital distance is the *mean matrimonial radius* that uses the place where the marriage partners were married, along with their two birthplaces, to define a "mariage triangle." The three distances are then averaged to calculate the mean matrimonial radius (Salzano and Freire-Maia, 1970). F. Salzano and N. Freire-Maia (1970) suggest that calculation of the measure can be simplified by (1) considering distances within a locality and its immediate hinterland as *zero* (i.e., if marriage partners were born and married in different parts of the same city, the Mean Matrimonial Radius is defined as zero); (2) distances between localities are assumed to be straight-line distances, even though this will always introduce a certain amount of error (the reader should note, however, that under certain circumstances such an assumption will introduce so much error that it cannot be used); and (3) if a locality cannot be precisely located on a map, the nearest county seat or state capitol is used.

The distribution of marital distances within a population is a useful indicator of marriage patterns in space. Simple bar graphs or histograms are constructed by plotting the distance between the birthplaces of marriage partners on the horizontal axis against the percentage of marriage partners in the population contracted at each distance on the vertical axis. Rare marriages contracted between partners born great distances apart are commonly grouped together in a single class. For example, if most marriages occur between individuals born within 100 miles of each other, but a few marriages are contracted between 100 and 5000 miles, the last class should group together all marriages between 100 and 5000 miles, regardless of the scale used up to that point.

Postmarital distance is a measure of where, geographically, offspring are born in relation to the parent's birthplace. It is calculated by averaging the lengths (in miles or kilometers) of the sides of a triangle defined by the two birthplaces of the parents and the birthplace of their *last born offspring* (Harrison and Boyce, 1972, p. 132). Histograms can then be constructed showing the distribution of postmarital distances for the entire population.

POPULATION SIZE

Abundance is one of the most obvious variables of a population. It is measured either by total *size,* the absolute number of individuals in the population, or by *density,* the number of individuals per acre, square mile, or some other geographical unit. Density is particularly useful because it can measure abundance variability within the population. Nevertheless, it can be calculated in several ways, including (1) the number of individuals per unit of total space, or the *crude* density, and (2) the number of individuals per unit of habitable space. The latter, known as *economic* or *ecologic* density, takes into consideration that not all parts of a geographical area can be lived in. For example, human populations are seldom inclined to live on water surfaces.

Collecting reliable data on abundance is a difficult task. The most accurate method is to directly count each individual, an approach used by ecologists working with territorial birds and by the United States national census. Even direct counting is subject to considerable error. National censuses, for example, have been plagued by intentional misreportings, particularly if the census is used for purposes of taxation or military conscription. Direct counting is also expensive and time-consuming so that a method of *sampling* is more commonly used. In anthropological studies this means that a few groups assumed to be representative of the population as a whole are selected and counted. Total population size or density is then inferred from the "samples." Sampling methods are less reliable because not only are the problems of direct counting still there, but also errors due to inference from the sample to the population are introduced. A small sample size is especially troublesome.

The relative abundance of populations has traditionally been used as a measure of adaptive success. Thus hunting and gathering populations with a density of one person per square mile are considered to be less successful than advanced farmers with densities of 10 to 200 persons per square mile. It would seem obvious that a large population has solved the problems of survival better than a small one. Yet population size or density cannot be used as a measure of adaptation without qualification since, for one reason, abundance is subject to drastic change over time. A large population may be able to exist only temporarily, "crashing" to a much smaller size in the long run because it has overexploited food resources, has been wiped out by an epidemic disease, or has violated some other ecological limit. Furthermore, abundance is only *one* measure of adaptation, as was discussed in Chapter 2. In general the problems underlying our understanding of the relationship between population abundance and adaptation are tied to the concept of carrying capacity and will be considered in detail in Chapter 11.

Aside from its use as a measure of human adaptation, anthropologists have been interested in population abundance as a possible cause of diversity in social organization. Moni Nag (1973) has recently reviewed the relevant literature and finds that kinship, social complexity, social stratification, and political organization have been proposed to be closely correlated with population size or density. Julian Steward was probably the first to see a relationship between kinship and population size. Patrilineality of the Western Shoshoni, according to Steward (1955, pp. 123–135), is one way of integrating widely dispersed nuclear families into a band, although a loosely knit one. Low population density and size have been invoked to explain the origin of the Australian sectional system of marriage (Yengoyan, 1968). The section system is found in the interior deserts of Australia, an area dominated by low rainfall and low population densities. Sections have often been equated with marriage classes, as a way of regulating who marries whom, but they are better defined as categories combining kin relationships for several purposes, including marriage, ritual, and subsistence. In addition, sections have territorial boundaries corresponding to the boundaries of the local groups included in each section. The number of sections included in a marriage system, then, is simultaneously a measure of the number of eligible mates, the number of local groups with which one has kinship connections, and the geographical extent of one's kinship connections. Increasing the number of sections in a habitat where population density is low and where local groups are small correspondingly increases the number of persons eligible for marriage. At the same time increasing the number of sections also provides a mechanism for "leveling out" local fluctuations in resources. That is, if resources suddenly fail in one locality because of drought, sections in other localities can be visited and called upon for economic support. Consequently, the largest number of sections is expected in habitats with the lowest population densities and group sizes and with fluctuating resources, while the smallest number is expected in habitats with the highest population densities and group sizes and with stable resources.

The origin of larger kinship groupings has also been attributed to population size. For example, in his classic study of the social organization of the western Pueblos of the American Southwest, Fred Eggan (1950, pp. 121, 288, 300) sees the development of both multilineage clans and phratries as responses to population growth, as well as the abandonment of an early system of cross-cousin marriage. Both organizations are better able to handle large populations and to protect the clan from extinction. Somewhat earlier Walter Goldschmidt (1948) argued that in aboriginal California the clan itself was a logical response to increased population size.

Population pressure, a concept that is more in vogue presently but often used synonomously with large population size, has been used to explain the origin of kinship organizations in a few studies. Thus Robin Fox (1967, pp. 152–153) contends that cognatic descent groups are likely to be adopted under conditions of high population pressure. According to Fox the advantage of cognatic descent is its flexibility. At any time some members of a group will be members of other groups as well. If that group becomes too large for the amount of land or resources available, then those members can leave to join other groups and take up land and resource rights in the new location. In such a way a population can be redistributed and population pressure reduced. By contrast, a unilineal descent group has no such flexibiltiy and, if the group becomes too large, pressure continues to build and eventually may cause the disintegration of the group. Michael Harner (1970) disagrees, at least for farmers. Harner postulates that cognatic descent is typical of farmers at low population densities, but as population pressure increases

subsistence resource land will become scarcer and hence more valuable, leading to competition for its control. Competition for scarce subsistence resources will, in turn, lead to ever larger local and inter-local cooperative social units to ensure success in holding and acquiring scarce resources. Typologically this will evince itself as a gradual shift from cognatic kinship to lineages. From lineages there will occur a shift to larger alliances in the form of non-localized clans or sibs, to facilitate multi-community organization for cooperative competition, particularly for competition in the form of war. (1970, p. 69)

A fundamental study by Raul Naroll (1956) investigated the relationship between *complexity* in social organization and population size. Population size is measured by "the number of people in the most populous building cluster of the ethnic unit studied (Naroll, 1956, p. 692)." The number of occupations, measured by the number of craft specializations, is used as one indicator of complexity in social organization. Complexity is also measured by the number of social units that link together individuals into an interdependent, symbiotic network. Each social unit is called a "team" and is defined as "a group of at least three people with clearly defined membership and formal leadership in regular use (Naroll, 1956, p. 696)." In effect these two indicators of complexity measure the degree of functional specialization and the degree of symbiotization. Naroll concludes from this study that there is a general trend toward increasing specialization and symbiotization in social organization as population size increases.

In a similar vein but with a different measure of complexity, Robert Carneiro (1967) reaches the same conclusion. He uses an index of social complexity based upon the presence or absence of "organizational" traits, selected from a list of 205 including craft and service specializations, merchants, markets, taxation, slavery, towns, cities, codes of laws, and a hierarchy of priests. Population size is shown to be directly correlated with the index. However, Carneiro cautions that increasing population size is not the only cause of organizational complexity and mentions the role of warfare as an alternative. Furthermore, increasing population size does not always result in organizational elaboration; the excess population may simply split off from the main body and move elsewhere. In later studies Carneiro (e.g., 1970) outlines the conditions under which population increase is likely to cause an increase in social complexity —namely, geographical and social circumscription.

Population density is suggested by Morton Fried (1967) to be an important cause of *social stratification* and *political centralization*. Basing his conclusions on a review of the relevant literature, Fried feels that increasing population density leads to difficulty in obtaining access to resources and social groups with limited membership will evolve to hold and manage such resources. He follows other authorities in rejecting the earlier view held by Meyer Fortes and E. E. Evans-Pritchard that complex social and political institutions are not found in those societies with the greatest population density (e.g., Stevenson, 1968). Fried accepts the thesis that population density is not the only cause of complex societies, that it is probably part of a multiple-causative network, but believes that "additional causal factors, when identified, will be difficult to isolate since I expect them to be capable, at least in part, of being represented as aspects of basic demographic problems." (1967, p. 204).

Kent Flannery (1969) has argued for a similar relationship between population density and complex societies in western Iran. Increasing population density stimulated the shift from hunting of small ungulates (e.g., deer) during the Upper Paleolithic to dry farming sometime around 10,000 to 7500 BP and then to irrigation farming between 7500 and 5000 BP. According to Flannery the changes in the intensity of land use were sufficient to set into motion processes culminating in complex social and political institutions such as the state. The primary reason is that as land is used more intensely, there is not only an increasingly large gap between the most valuable land and the least valuable land but also a progressively smaller amount of valuable land. Consequently, specialized social groups are more likely to appear for the purpose of holding and managing these scarce resources, a suggestion following that of Fried. For example, nearly 35 percent of the total land of western Iran is optimal for the hunting of ungulates. However, the shift to dry farming dropped the percentage of optimal land to 10 percent and the shift to irrigation farming to only 1 percent! Today 30 percent of the total farming yield comes from this small amount of land, yet it is owned and controlled by only 1 percent of the population.

AGE AND SEX STRUCTURE

No population is made up of individuals who have the same age and sex. Also, the actual ages and sexes, the age and sex structure, vary from one population to another. The differences are caused by variation in the rate at which members are being added or lost to each age and sex group, changes that are sometimes the result of ecological processes.

Age Structure

The age structure of a population is described by age intervals and age pyramids. An age interval is an arbitrarily selected range of ages within which everyone is assumed to have identical demographic characteristics, including birth rate, death rate, expected age of death, and so forth. The smaller the range of ages, the more accurate is the assumption.

For most purposes a 5-year interval is sufficient, grouping individuals into such categories as 0 to 4 years, 5 to 9 years, 10 to 14 years, and so forth. This assumption is not always warranted, and it may be necessary to reduce the size of the interval to achieve greater accuracy. On the other hand, poor demographic data, which must often be used by anthropologists, may require increasing the size of the interval to eliminate many uncertainties about the age of individuals.

The age pyramid is used to graphically depict the age structure of a population. Along the vertical axis of the pyramid are arranged the age intervals, with 0 to 4 years at the bottom and oldest age interval at the top. The pyramid is split right down the middle by the horizontal axis. The left side shows the number of males in the population and the right side the number of females. The horizontal axis is scaled so that 0 corresponds to the pyramid split and values increases both to the left and to the right. For each age interval it is then possible to draw a bar or rectangle whose length to the left of 0 represents the proportion of males in the population and whose length to the right represents the proportion of females. Correct scaling of the vertical and horizontal axes of the pyramid improves its readibility. Demographers (e.g., Pressat, 1972) suggest that the vertical axis of a pyramid should have a height equal to two-thirds of the *total* length of the horizontal axis (i.e., the combined lengths of the male and female segments).

The shape of age pyramids is an important indicator of the way in which populations are growing. Figure 7.2 illustrates the age pyramids

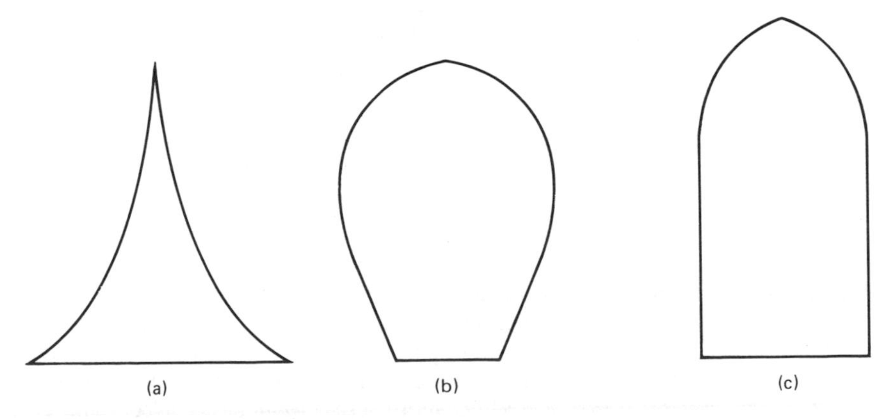

FIGURE 7.2 Age pyramids: (a) a bottom-heavy pyramid with a large proportion of young individuals, (b) a top-heavy pyramid with a large proportion of old individuals, and (c) a stationary pyramid with nearly equal proportions of individuals of all ages.

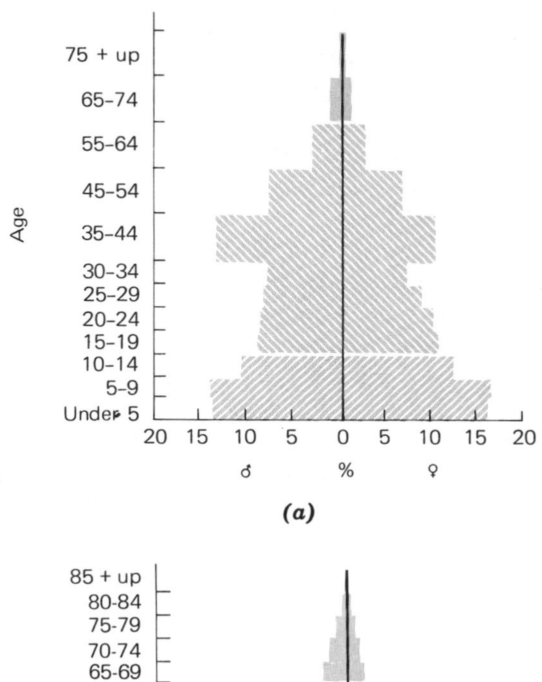

(a)

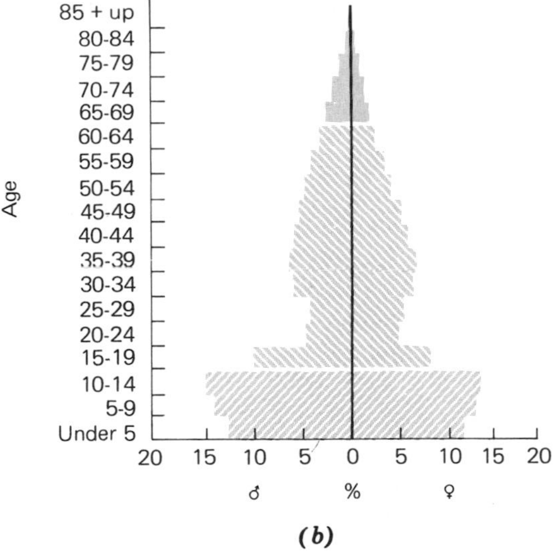

(b)

FIGURE 7.3 The impact of migration on the age structure of Appalachia. (a) In 1930 people were moving into appalachia in search of mining jobs. (b) in 1960 young people were leaving the economically depressed appalachian region for better jobs in large eastern cities. (From Robert Leo Smith, *Ecology and Field Biology,* 2nd ed., New York: Harper & Row, 1974. Copyright © 1974 by Robert Leo Smith. Reprinted by permission of the publisher.)

for three hypothetical populations. The first is bottom-heavy and reflects a relatively large proportion of younger-age classes in the population and a high birth rate. Populations with a bottom-heavy age structure have a tremendous potential for growth if death in the younger age classes is not high. The second age pyramid is top heavy, reflecting a relatively large proportion of older-age classes. Top-heavy populations are declining in size, since there are not enough young to replace those dying off at the top. Finally, the third age pyramid is stationary and gives zero population growth. Such populations have nearly equal proportions of all age classes with the exception of the very oldest, where rapidly increasing death rates bring the age pyramid to a point.

Births and deaths are not the only events affecting the shape of age pyramids. Migration into and out of the population is equally important. The impact of population movement on Appalachia is illustrated in Figure 7.3. Pyramid 1 shows the 1930 population and reflects the immigration of people into the mining region. Pyramid 2 shows the 1960 population and reflects the emigration of young people to urban areas. Colonization, of all the different kinds of migration, has the greatest impact upon population structure. Age structure is usually heavily biased in favor of young adults, and males typically dominate the sex structure.

Sex Structure

Sex structure, the proportion of males and females, is another population characteristic that is suspected to have ecological relationships. The sex ratio is a useful measure of the sex structure in a population and is calculated simply by dividing the number of females in the age group (or the entire population) into the number of males and multiplying the result by 100. That is, (males/females) times 100. Sexual "balance" is indicated by a sex ratio of 100, a value possible only when the number of males and females is equal. A high sex ratio (over 100) indicates an excess of males over females, and a low sex ratio of under 100 shows relatively more females in the population. In humans the general sex ratio for large populations is supposed to be 105 at the time of birth. That is, there are slightly more males than females born. Why? One possibility is that more sperm that carry the y or male-determining chromosome are produced or survive to reach the ovum (Teitelbaum, 1972). However, in small populations *stochastic* (that is, random or unsystematic) fluctuations in reproduction often result in sex ratios that are considerably higher or lower than 105. Cultural practices after birth, such as *female infanticide* (intentional killing of female infants), change sex ratios even more. Thus Joseph

Birdsell (1975, pp. 367–71) has collected data showing that the aboriginal populations of Australia had an average sex ratio of 153 among young children. The large excess of males he attributes to the practice of female infanticide.

In subhuman primates sex ratios have been observed to be greatly biased toward females in arid deserts with scarce, dispersed food resources. For example, the ethologist Hans Kummer (1971) documents a "harem" organization in troops of Hamadryas baboons in the arid desert of Ethiopia, harems consisting of a single adult male with several adult females and their immature young. Why? It has been suggested that a low sex ratio is adaptive in habitats with a low-energy content (Denham, 1971). A single male is capable of impregnating several females and assuring the reproductive success of the population. Other males are reproductively unnecessary (unless the females can afford to keep them as pets). The scarcity of food places a premium upon the smallest number of individuals that can still maintain reproductive viability and, therefore, a low sex ratio is adaptive.

Conversely, a high sex ratio is adaptive in populations with a strong hunting focus and a chronic scarcity of food. The Polar Eskimo is the classic example. Nearly all of the food supply comes from the hunting of a variety of large land and sea mammals; however, the energy content of the region is so low that starvation is frequent, and there is strong selection to keep the population as small as possible. Since males do the hunting and provide nearly all the food consumed, females are "surplus" and are systematically removed from the population through the practice of female infanticide. Consequently, the sex ratio is high during much of the reproductive period but eventually balances because of a high male mortality rate.

Sex ratios can also be changed dramatically by hazards. Human populations with an ocean fishing focus are notably subject to disasters caused by "swamping" of fishing boats on the high seas and drowning of the male fisherman. Anyone who has visited the coastal fishing villages of Portugal, for example, will be immediately struck by the large number of widows of dead fishermen in the community and the relatively small number of adult males. A specific example of the impact of a boating accident upon the sex ratio of human population comes from the South Atlantic island of Tristan da Cunha, a former British garrison during the Napoleonic War (as described in Roberts, 1968). After the end of the war a few men stationed at the garrison continued to remain on the island, along with their wives, and founded a rather unique group. From an initial size of 15 in 1816, it grew to 103, but subsequent loss of group cohesion caused by the death of the founder led to the out-migration of a large part of the population between 1855 and 1857. However, the population recovered and once again increased to 106 by 1885. On November 28, 1885, a sea disaster drastically reduced the size of the population, this time strongly affecting the sex ratio. The island has no natural harbor and, when ocean-going vessels stopped, the islanders sometimes went out to meet them in small boats for purposes of trade. On this particular day a boat with 15 adult males set out to meet a passing vessel but suddenly sank, drowning all of the occupants. According to D. F. Roberts, who has studied the island population extensively, the disaster turned Tristan into an "island of widows," leaving only four adult males (1968). Although during the next few years some of the widows left the island, the sex ratio of the population was significantly affected for some time.

Relationships with the "social" environment can have an equally dramatic impact upon sex structure. Data for the Yanomamo, tribal farmers in Venezuela and Brazil, show that warfare between village populations is one of the most important causes of death (Chagnon, 1974, Table 4-10). In two villages the ethnographer recorded 110 deaths due to warfare; of these 96 are males and only 14 are females. The impact is great enough to cause a large sex imbalance in a small population, an impact that has a number of social consequences.

DISTRIBUTION

All populations have a geographical distribution that can be partly explained by ecological relationships. The distribution of a population can be described by a set of points plotted on a map; each point represents an individual, a house, a camp, a town, or some other distinct entity. Numerical techniques are then used to decide if the distribution of points is random and, if not, what kind of nonrandom distribution it most closely approximates. Finally, models are constructed to explain the distribution.

Sampling Problems

The explanation of a population's distribution, ecological or otherwise, depends upon accurate data. Yet, as pointed out earlier, most demographic data are not complete but are *samples;* whether or not the sample is representative of the actual distribution is a real problem. In anthropology the problem has been that demographic data are often collected not as a major goal but as a minor part of a field study focusing on something else. As a result the data collected on how humans are distributed reflect the interest of the field worker in another problem; they do not reflect reality. The bias introduced by this "haphazard" method is clearly illustrated by early settlement pattern studies in archaeology. Large, spectacular sites were chosed for research purposes while small outlying sites were ignored. This approach, of course, reflected not so much a weakness in archaeological method as it did an overriding concern with problems that were not demographic. With the advent of more demographically oriented research problems in the early 1950s archaeological cultures such as the Adena in the Ohio Valley and the Maya of Mexico and Guatemala appeared to have ceremonial centers with few residents and a very low population density elsewhere. The question of how such centers could arise in such a low population area naturally appeared. However, in the last few years work on the small "insignificant" sites in this area and in the Mayan area has shown a very different picture. At Tikal, for example, a quite large residential population has been postulated at the periphery of the ceremonial center.

Recent concern with demographic problems in anthropology has brought about a "revolution" in the methods used, or at least proposed, to collect demographic data. The "new methodology" is in large part adopted from disciplines with a longer history of interest in distributional problems, for example, plant geography, and much of it has to do with methods of sampling that are appropriate for reliable statistical inference. A detailed discussion of sampling is beyond the scope of this book, and the interested reader can consult a number of recent publications in this area. However, the principal problem areas are discussed in the following paragraphs (based on Krumbein, 1965; and Mueller, 1974).

There are several ways of taking samples from a complete population, all of which give each member of the population a chance of being selected. Exactly *what* chance varies from one sampling plan to another. In *simple random sampling* all individuals (person, household, settlement, or other "sampling unit") in the population are given an *equal* chance; the method is to select as many individuals or sampling units as necessary by means of a table of random numbers or other randomizing device. For example, suppose that we were interested in the distribution of household size in a community. Simple random sampling would require giving each of the households a different number, say 0 to 100, and selecting several numbers randomly. One problem with this plan, of course, is that all of the numbers selected may belong to households adjacent to each other, preventing the sample from reflecting the composition of households in other parts of the community. *Systematic sampling* is one way of preventing this problem. In this case the community is arbitrarily divided into sections that do not overlap, each of which contains a given number of households. The households in each section are numbered consecutively in the same way, for example, 0 to 9. Then the number of a household is selected randomly from the first section, and the same number is used to select a household from each of the remaining sections. In this way the household sample is evenly spread out over the community, and a better idea of the entire population can be had. Forced spacing of the sample, however, introduces other problems, particularly reliability in estimates of population variability. *Stratified sampling* is a way of improving variability estimates while keeping a good coverage of the entire population. The population is divided into *sampling strata* using arbitrary or "natural" criteria (e.g., land use, rainfall zones, economic zones). Simple random samples are then taken from each stratum. The use of sampling strata spreads out the sample, although not as much as in systematic sampling, and allows reliable studies of the variability within each stratum. The fourth common sampling strategy is *cluster sampling*. Using this approach, the investigator randomly selects a sampling unit and then includes this unit *plus* one or more adjacent units in the sample. That is, a *cluster* of units, not just one, is taken at each sampling location. Cluster sampling can be used in conjunction with other sampling strategies, for instance, by taking clusters within sampling strata.

How large the sample should be depends upon many factors, including the desired degree of reliability, the kind of data being collected, the kind of problem being studied, and the sampling strategy being used. In general the absolute size of a sample is more important to statistical inference than is its relative size expressed as a proportion of the total population. There are mathematical methods for calculating exactly how large a sample should be for any particular degree of accuracy, and any textbook in statistics can be consulted for this information.

The *size* and *kind* of sampling unit are as essential as sample size to the reliability of distributional studies. Consider the quadrat, a commonly used sampling unit. The quadrat is traditionally a square-sided plot, although the term is often used to refer to circular plots, rectangular plots, and a wide variety of other shapes. The actual shape of the quadrat has been shown to affect results; thus in vegetation studies "rectangular units are more efficient than square plots of equal size. Short strips give less variable results than do squares but short strips are more variable than long strips" (de Vos and Mosby, 1969, p. 156). Quadrat size also affects results, and the smallest size that will give reliable results must be worked out in the field. One way to do this is to "nest" quadrats of different sizes (i.e., place smaller quadrats within larger ones), keep the data separate for each size quadrat (e.g., the percentage of each kind of settlement showing up), and plot the quadrat size against the data. The point at

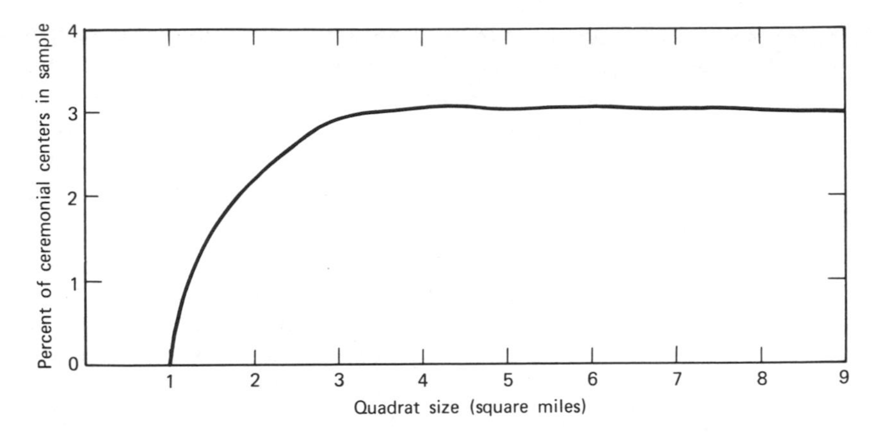

FIGURE 7.4 Estimating minimal sampling units. Quadrat sizes are plotted on the horizontal axis. Data from each size quadrat are plotted on the vertical axis. In this case the percentage of ceremonial centers showing up in the human settlements found in each quadrat are being plotted. The point at which the curve flattens out is the smallest size that can be reliably used, in this case about 3 square miles. (The method is discussed in de Vos and Mosby, 1969.)

which the curve flattens out is the smallest size that can be reliably used (de Vos and Mosby, 1969, p. 156). Figure 7.4 illustrates the method.

Other sampling units, such as *transects*, are also commonly used, and the investigator will have to make tests to decide whether quadrats, transects, or other sampling units are best suited for the problem at hand.

Distribution Models

Models explaining the relationship between the distribution of a human population and the environment are numerous but are generally of four kinds: *barrier, gradient, locational,* and *social or cultural*.

BARRIER MODELS. Barrier models explain distribution by the presence of obstacles in the environment. The enforced separation of populations brought about by barriers, for example, is an important part of a recent model of human evolution suggested by Adriaan Kortlandt. According to Kortlandt, *it should be possible . . . to reconstruct phylogenetic trees of descent from paleogeographical and paleoecological maps that indicate the relevant faunal barriers. Neither the apes nor man can swim by nature; nor can the apes carry food and drinking water (juicy fruit) over long distances. Thus we have to look for water barriers and arid zones to explain and reconstruct the speciation and divergence of apes and hominids. The core point is that neither incipient man nor the various ape species could ever have branched off as long as they enjoyed free sexual intercourse with one another. (1974, p. 427)*

Physical barriers such as deserts, large bodies of water, and glaciers are not the only obstacles to human movement. Biological barriers (zones of epidemic disease, for example) and cultural barriers (such as zones of intense warfare and "sacred" places) are also important.

GRADIENT MODELS. The gradient model assumes that continuous gradations in the environment, such as temperature, altitude, or annual rainfall, are correlated with the distribution of a population. Since the problems of survival are somewhat different from one gradation to another, a population's distribution along a gradient reflects the conditions under which it can best survive and therefore obtain its ecological niche. This says no more than that some populations are better adapted to, for example, high annual rainfall while others are better adapted to low annual rainfall. In an effort to separate distinct life styles, ecologists often arrange populations along an environmental gradient to see if they are adapted to the same or different gradations, a technique called *ordination*. This is, of course, the same method often used to define ecological niches.

LOCATIONAL MODELS. Locational models have long been used by economic geographers to explain the geographical distribution of human settlements in market economies. It has not been until recently that they have been adopted by anthropologists. The basic assumption of locational models is that human beings are economically centered and that their behavior is adequately described by formal economic theory. Accordingly, human settlements are situated so as to *minimize* the effort (or cost) in the production and distribution of needs and wants. The "economic man" concept was developed in Western market economies and, consequently, locational models have been most frequently applied in this context. However, at least some economic anthropologists now see the "economic man" concept as applicable to all human groups, suggesting that locational models may have wide application (e.g., Schneider, 1974).

The "central place" model is one kind of locational model that has been used by anthropologists in a number of situations. For example, the archaeologist Gregory Johnson (1972) applied the central place model to the study of early settlement patterns in Mesopotamia. According to Johnson, the model assumes that human settlements of about the same size in the most efficient ("least expensive") exchange network will be situated equidistant from each other in a hexagonal lattice. Each settlement is a central place or exchange node for a surrounding region of hexagonal shape. Within each of these hexagonal service "cells" is a nested hierarchy of progressively smaller central places. Johnson uses large towns, towns, large villages, villages, and hamlets to make up the hierarchy. Thus a large town is a central place for the production and distribution of goods and services to surrounding towns; a town is a central place for servicing surrounding large villages; and so forth. Based on these assumptions the central place model is then used to construct an expected distribution of settlements in the Diyala Basin in Mesopotamia, which is then compared to the actual settlement distribution. According to Carol Smith, deviations from the settlement distribution expected by the central place model arise from violation of one or more of the following assumptions:

1. Market exchange is integrated and part of a regional system.

2. Market centers exist to facilitate market exchange and are located to minimize the effects of distances.

3. The landscape is featureless, and transportation ease is the same in all directions.

4. The population is evenly distributed in the marketing region.

5. Market suppliers are seeking to maximize profits and market consumers are seeking to minimize costs.

6. Market suppliers are competitive (1974, 168–69).

SOCIAL OR CULTURAL MODELS. Social models explain the distribution of human populations as an adaptation to social and cultural environments. The movement of Jews to Israel, Mormons to Salt

Lake City, and Muslims to Mecca, for example, has nothing to do with either physical geography or the principle of least cost; it has a lot to do with ideology. Similarly, the movement of political refugees to countries more amenable to their political beliefs can be explained only by social and cultural variables. The concept of *social space* or *social distance* is important to social models. Social space is measured by the intensity (or frequency) of social interaction between individuals, based upon physical proximity, participation in exchange networks, marriage alliances, and ritual partnerships, among other things. Populations that are close together in social space can exchange members readily and, as a result, their composition is subject to rapid and drastic change as circumstances dictate. The tightening of social boundaries sharply reduces exchange of members and is often associated with conflict. The definition of the population itself sometimes depends upon proximity in social space rather than geographical space. The migration of individuals from their homeland to nearby cities, other countries, and all parts of the world creates widely dispersed enclaves or colonies. The migrant colonies often continue to maintain close ties with their homeland and, indeed, may support it economically by sending back home a large part of the wages that are earned. In many cases migrants prefer to marry persons from their homeland rather than from regions that are geographically closer. The population is thus not a group of individuals who live together and are in frequent contact with each other. Economically and ecologically, the population is made up of widely dispersed enclaves.

Transhumance

The distribution of human populations can also be explained by seasonal movement from one ecological zone to another, a practice called *transhumance.* Fredrik Barth (1961) gives an example from southwestern Iran. The Basseri are pastoral nomads who depend mostly upon herding of sheep and goats and who occupy a hot, arid, and diverse habitat ranging in elevation from around 2000 to 13,000 feet. Pastureland for grazing the herds is essential to the Basseri life style, and a pattern of transhumance is followed to take advantage of seasonally available pastures at different elevations. In the winter Basseri camps are found in the southern part of their habitat at low elevations. Winter pastures in this area are poor but extensive, while the pastures further north are at higher elevations and covered with snow. During this time the Basseri are camped on mountain flanks or ridges to avoid mud and flooding. Migrations from winter camps are local and short. As spring progresses the pastures at low and middle elevations begin to dry up and, at spring equinox, the main migration to the northern mountains is started. During this time the camps are moved nearly every day. In the summer Basseri camps are found in the northern pastures at high elevations above 6000 feet; they are relatively sedentary, moving infrequently and locally. By August the camps begin the return trip south. Fall is generally a poor time for pastures everywhere, because of drought, but the Basseri herds feed upon the stubble of farmlands that have just been harvested. During this period, as in the spring, camps are moved often. With the advent of winter the Basseri are back in the southern pastures and the cycle begins anew.

SUMMARY

Population studies are useful to the ecological anthropologist for several reasons. The population is especially useful in the study of adaptation. Abundance, age and sex structure, and distribution are three population variables that are especially subject to adaptive change. The success of a human group in coping with environmental problems is measured, at least indirectly, by population size or density. There is also evidence that some variability in social organization is closely correlated, perhaps causally, with variation in abundance. Sex ratio, a measure of the male/female composition of a population, often reflects hazards in the environment, including warfare and hunting accidents, as well as adaptive behavior. Finally, the distribution of a human population varies because of ecological relationships with the physical, biological, social, and cultural environments.

EIGHT POPULATION INTERACTION

As suggested in the preceding chapter, many of the problems that a human population must face in its environment have nothing to do with the landscape or climate. Instead they are caused by interaction with other organisms, some of which are human and some of which are not. The kinds of interaction taking place are numerous but are usually grouped into three categories: neutral, negative, and positive (E. Odum, 1971, p. 211). *Neutral interaction* results in no effect whatsoever, neither beneficial nor harmful, on any of the participating populations. Interaction causing definite harm to one or more of the participants, *negative interaction,* is considerably more common. *Competition* is the striving of populations for limited needs or wants (e.g., food, space, prestige items). One kind of competition is "direct interference" in which one or more of the participants directly interfere with the accessibility of the others to limited needs or wants. Concepts of territoriality and tight social boundaries underlie interference competition. The other kind of competition is "scrambling" in which the participants consume limited needs and wants without directly interfering with each other. Loose social boundaries are associated with scrambling competition, and it is typically found among mobile hunters and gatherers.

There are three outcomes to competition: replacement, unstable coexistence, or stable coexistence (R. Smith, 1974, p. 402). The most common outcome is the replacement of one or more of the competitors by the others. Some years ago the biologist Garrett Hardin (1960) referred to this process as the "competitive exclusion principle," building on earlier mathematical theory and experimental data. In effect the principle states that two populations cannot permanently occupy the same ecological niche. One of the populations will either become extinct, leave the habitat, or change its life style. During the process of exclusion, however, the competing populations oscillate in size and establish a temporary, unstable equilibrium with each other. The concept of *ecological overlap* is important in understanding the relationship (R. Smith, 1974, pp. 402–403). Any population tolerates a variety of environmental conditions but some habitats are *optimum* for survival while others are *marginal*. At a small size the population occupies only the optimum habitat but with increasing size expands into the marginal area. However, other adjacent populations with similar life styles may also occupy the marginal habitat. The area held in common is the ecological overlap. The relationship between the two depends upon the strength of the competition. If the expanding population encounters strong opposition in the area of overlap, it will be forced to remain in the optimum zone. However, when its competitor is absent from the marginal zone, it can once again expand. The final outcome of this kind of competition is replacement. Under some circumstances competing populations having the same life styles can coexist permanently. If the competitors have a low population density in relation to the resources available in the habitat or if they both have highly generalized requirements, they can occupy the same area indefinitely. For that reason the competitive exclusion principle may not be applicable to the interaction between many human populations, particularly generalized hunters and gatherers.

The other kinds of negative interaction are predation, parasitism, and antibiosis or amensalism (E. Odum, 1971, p. 211). *Predation* is the exploitation of one population by another, resulting in the transfer of materials and energy from one trophic level to another. Predator-prey relationships are often complex and important to the regulation of the population size of both the eater and the eaten. *Parasitism* is the coexistence of two populations under conditions in which one "taps" the other for energy and materials. Unlike the predator the parasite does not kill its prey. Indeed, ideally the parasite's host is affected only slightly, as is true, for example, of the relationship between German measles and man. Finally, *antibiosis* is a kind of interaction resulting in definite harm to one population and in no benefit or harm to the other. The usual example is the population that releases a harmful substance into its environment.

Positive interaction includes commensalism, cooperation, and mutualism (E. Odum, 1971, p. 211). *Commensalism* is a relationship that is beneficial to one of the participants and neither good nor bad to the other. Barnacles and orchids, for example, depend upon attachments to other organisms but do not affect them. *Cooperation* is any kind of positive interaction in which both populations are mutually benefited. For example, in East Africa baboons and gazelles are common companions to provide mutual protection from predators, the relationship founded upon the good vision of the baboons and the good hearing of the gazelles.

Cooperation among human populations freqently involves the exchange of needed or wanted resources. *Mutualism* is cooperation taken one step further, to the point that the survivial of the populations is completely dependent upon their cooperation. The integration of economically specialized human populations into exchange systems is an example of mutualistic interaction. One cannot exist without the other.

INTERACTION BETWEEN HUMAN AND NONHUMAN POPULATIONS

Human populations enter into many kinds of relationships with nonhuman populations, including predation, parasitism, and cooperation or mutualism. Predation is necessarily an important interaction for any consumer organism because food is obtained in this way. Man is no exception. Of course, humans are not always on the predator end of the relationship, and some authorities have argued that early hominids such as the australopithecines were subject to rather intense predation from large cats and other carnivores. Nevertheless, there is little evidence that nonhuman predators have played any role at all in recent human history.

It is often difficult to decide whether cooperating populations are dependent or not upon the relationship, so for our purposes cooperation and mutualism will be treated as a single category. Cooperative interaction between human and nonhuman populations is illustrated by the relationship between wild dingoes in Australia and the Walbiri (Meggitt, 1965). Walbiri hunters would find and follow the fresh tracks of dingoes chasing a kangaroo. With luck the dingoes would be found just as they were attacking the kangaroo, a dangerous job at best. The Walbiri hunters would then kill the kangaroo without hurting the dingoes and butcher it on the spot. Most of the meat was taken back to the Walbiri camp, but scraps were left for the dingoes to consume. The relationship between domesticated plants and animals and man is a more common form of cooperation. In many cases clear mutualism is involved, as is true, for example, of the relationship between corn and the corn farmer.

Humans act as hosts to numerous parasitic microorganisms, a relationship often identified with human diseases. Parasitism is one of the most important aspects of human ecology although it seldom receives the attention it deserves in anthropological textbooks. Consequently, the subject is treated in some detail in this section. Disease microorganisms affect human populations in a way determined by demographic, physical, and cultural factors. The most important of these are (numbers 1–4 from Alland, 1970):

1. Human population size.
2. Human population distribution.
3. The isolation of social groups.
4. Patterns of human movement.
5. Ecological complexity of the habitat.
6. Unique cultural patterns.

The interaction between disease microorganisms and human population size is illustrated by measles. Measles is a communicable disease that is highly contagious for a few days, but infection of an individual is followed by lifetime immunity. Consequently, when measles is introduced into a population, it spreads rapidly, reaches epidemic proportions, and then disappears. Even though immune individuals are gradually replaced by new susceptible individuals through birth and emigration, a measles epidemic will not strike again *unless* it is reintroduced from the outside or unless it persists permanently. It has been calculated that a minimum number of 4–5000 new individuals a year is required to permanently maintain measles in a population, a figure that is usually found only in populations of 300,000 or larger (Fenner, 1970, pp. 48–68). The relationship between population size and the persistence of measles is shown in Table 8.1. These data are taken from several island populations during the period from 1949–1964. Guam and Bermuda are exceptions to the trend, but both had significantly greater chances of reintroductions from the outside because of a constant influx of military personnel and tourists from the U.S. mainland.

The distribution of a human population often has a notable impact upon the persistence of a disease organism. Figure 8.1 shows the relationship between the duration of measles epidemics (measured in months) and the average distance between new births (who are susceptible to measles) in seven island populations of about the same size. The graph suggests that dense populations are struck with relatively short epidemics followed by rapid "fade-out" of the disease while dispersed populations increase the chances that an epidemic will survive for a long time.

The intensity of predation by disease organisms is sometimes affected by the way in which humans are distributed vertically. The hill North Vietnamese built their houses on stilts 8 to 10 feet above the

TABLE 8.1 PERSISTENCE OF MEASLES IN ISLANDS WITH POPULATIONS OF 500,000 OR LESS

Island	Population	Annual Births Less Infant Mortality	Percent Months with Measles
Hawaii	550,000	16,700	100
Fiji	346,000	13,400	64
Samoa	118,000	4,440	28
Solomon	110,000	4,060	32
French Polynesia	75,000	2,690	8
Guam	63,000	2,200	80
Tonga	57,000	2,040	12
Bermuda	41,000	1,130	51
Gilbert and Ellice	40,000	1,260	15
Cook	16,000	678	6
Falkland	2,500	43	0

Source. Frank Fenner, "The Effects of Changing Social Organization of the Infectious Diseases of Man." In S. V. Boyden, ed., *The Impact of Civilisation on the Biology of Man*, Toronto: University of Toronto Press, and Canberra: Australian National University Press, 1970.

ground, *above* the flight ceiling of the malaria vector *Anopheles minimus*. Consequently, even though malaria was endemic to the region, its incidence was kept at a low level by cultural adaptations. The hill North Vietnamese also housed their cattle in stables under the stilted dwellings, thus encouraging the mosquitos to bite the cattle rather than humans, and cooked within the houses, filling the living area with smoke and discouraging any mosquito that may have strayed beyond the normal flight ceiling. The *relocation* of human populations is frequently responsible for disrupting this kind of relationship. For example, Jacques May (1972, p. 25), a specialist in nutritional ecology and epidemiology, describes the impact of relocating lowland North Vietnamese into the hill area just described. The lowland Vietnamese are situated in the delta of the Red River and, unlike the hill area only 60 miles to the north, are not plagued by malaria-carrying mosquitos. One-story houses are constructed from mud and rice straw. Stables for domestic animals are found at the *side* of the dwellings, within the same vertical zone. Finally, cooking is done outside the house rather than in the living area. When the lowland populations were relocated to the hills because of population pressure in the delta, their cultural pattern was carried with them. The transplanted Vietnamese moved, in effect, into the vertical zone occupied by the malaria vector *Anopheles minimus* and the results have been disastrous: Malaria is decimating the migrants. May observes that the lowland Viet-

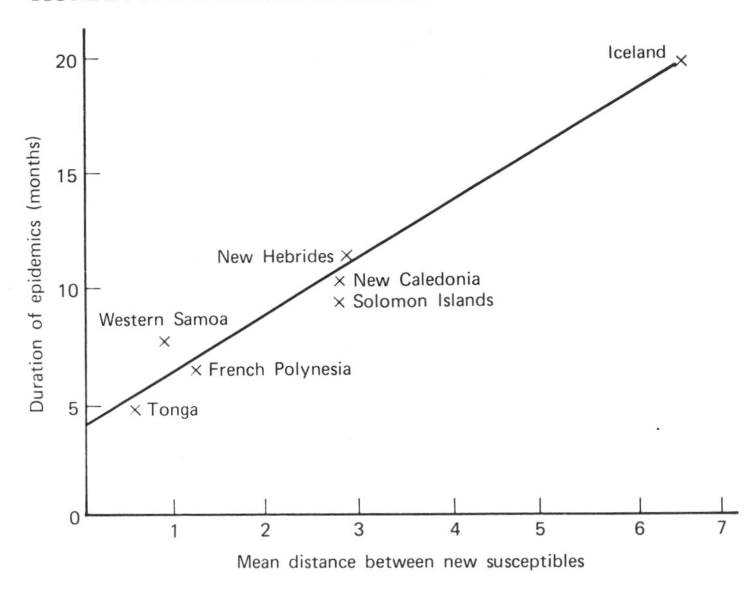

FIGURE 8.1 Population dispersal and persistence of measles in island populations. (From Frank Fenner, "The Effects of Changing Social Organization on the Infectious Diseases of Man," in S. V. Boyden, ed., *The Impact of Civilisation on the Biology of Man,* Toronto: University of Toronto Press and Canberra: Australian National University Press, 1970. Copyright © 1970 by the University of Toronto Press and the Australian National University Press. Reprinted by permission of the publisher.)

namese now view the hills as a region inhabited by evil spirits and feel that they should not go there, a belief that is consistent with the reality of malaria.

A similar incident has been documented in Africa. The Rukuba of northern Nigeria traditionally lived in easily defensible hilltop fortresses because of endemic warfare (Duggan, 1962). Although the settlements were well within a region of a sleeping sickness vector, the tsetse-fly *Glossina palpalis,* predation from the disease was not a problem because the settlements were situated above the flight ceiling of the fly. However, after the British established peace, the Rukuba began to move down from the hilltops into the fertile land below. New settlements were built directly in the fly zone, bringing humans and the tsetse fly into constant contact for the first time. All that remained to set into motion the sleeping sickness-fly-human cycle was the introduction of the trypanosome protozoa itself and that took place during 1931 and 1932. The ensuing epidemic of sleeping sickness decimated the Rukuba, forcing the abandonment of most of the valley settlements.

The rise of a new social group as a response to folk beliefs about leprosy has been documented among the Gojjam of the Ethiopian Highlands (Schofield, 1970). Here endemic leprosy is considered to be an *inherited* disease rather than a communicable disease. As a result marriage is strictly forbidden between individuals who have leprosy in their genealogy and those who do not. Leprous individuals and their offspring, therefore, belong to an endogamous social group. Members of this group are not kept isolated, however, since the disease is not believed to be communicable, and social mixing is commonplace without fear of being infected. Nearly 25 percent of the entire population is included in the leprosy group, and, interestingly enough, the Gojjam claim that no one born into the nonleprosy group later contracted the disease.

Although, in the above case, the rise of a new isolated group helped to contain an endemic disease, in other cases the existence of isolated social groups has encouraged the spread of an infectious disease. The relationship between smallpox and student priests in Ethiopia is a good example (Schofield, 1970). Every attempt is made to prevent individuals from leaving localities where smallpox has been reported. However, the success of this effort has been greatly hindered by the travels of student priests from one locality to another. Members of this group are particularly susceptible to smallpox because most come from rural regions where vaccination and treatment procedures are not used and because they believe that "worldly" care, including medical treatment, must be avoided. After a long childhood of studies in a rural priest-school, the student priests are expected to travel throughout the country for a year or more, living entirely from food obtained by begging or from other gifts of "divine" origin, before they can be ordained. Farmers, townspeople, and officials alike accept their philosophy and do nothing to prevent their travels or to influence them to accept medical care. As a result smallpox is often carried from one place to another by members of this group and reintroduced time and time again into local populations.

Because of the interaction between microorganisms and the distribution of human populations, one might suspect that human mobility also plays a role in disease. There is evidence that this is indeed the case. Hunting and gathering populations that are highly mobile are unaffected by the kinds of disease that thrive in accumulated human waste. Mobility not only means that little time is spent in any one place but also places strict limits on the amount of material baggage that can be transported from place to place. Both reduce waste and, therefore, a suitable habitat for many types of microorganisms. With increasing sedentism, however, waste accumulation becomes more and more of a problem. Microorganisms thrive and new diseases drastically alter the human condition. Epidemic diseases are particularly important and undoubtedly took a heavy toll in early villages, towns, and cities. However, before the development of effective transportation facilities, large-scale epidemics affecting many communities were probably uncommon because of little contact between populations.

In contemporary human populations that are highly sedentary, the existence of mobile subgroups often creates conditions suitable for the spread of microorganisms. For example, migrant laborers, commonly associated with economic development schemes, have played an important role in the transmission of sleeping sickness (trypanosomiasis) throughout many parts of Africa. During the period from 1920 into the 1940s, economic development in the Ashanti area of central Ghana attracted migrant laborers from northern Ghana and from the northern countries of what are now the Upper Volta Republic, Mali, and Niger (Scott, 1957). The migrants passed through the tsetse fly belt just north of the Ghana forest and were infected with the trypanosome protozoa carried by the tsetse fly. When the migrants reached the Ashanti country, previously free of endemic sleeping sickness, they introduced the disease. Ultimately, sleeping sickness became more common in the Ashanti country than in the traditional sleeping sickness belts further north. However, migrant workers often return to their homelands after several years of wage labor and, in this case, a serious outbreak of sleeping sickness was the result. Attempts to eradicate the tsetse flu in northern Ghana have been fairly successful; unfortunately, the return of the migrant workers continually reintroduces the trypanosomes, instigating new epidemics.

The intensity and kind of disease parasitism are closely related to the complexity and diversity of the ecological system within which humans participate. Hunters and gatherers, for example, occupy a range of environments from the complex and diverse tropical rain forests to the simple deserts; their disease characteristics vary accordingly. F.L. Dunn (1968) uses these two extremes as "ideal" types and outlines the implications of each for a population of hunters and gatherers. *Complex* ecological systems are marked by many different kinds of infectious and parasitic disease organisms, as well as a large number of species that can serve as vectors for the disease, as intermediate hosts for the temporary development of the disease organism outside humans, and as alternative hosts. As a result human populations in this type of environment have many different kinds of sexual and asexual infections (i.e., disease reproducing by sexual and asexual means). However, all species, including disease organisms, tend to have small population sizes in compex ecological systems. Consequently, the intensity of parasitism by each kind of disease is low. Furthermore, since disease vectors, such as mosquitos or flies, have low population numbers per species, their ability to transmit a disease to the human population is also low. At the opposite pole is the *simple* ecological system with few disease species, vectors, intermediate hosts, or alternative hosts. Sexual and asexual infections are few in kind, therefore, but the population size of each of these is large. For this reason the intensity of infection is high, as is the effective transmission of the disease through its vectors.

In 1956 a district medical officer, Dr. Vincent Vigas, discovered several cases of a previously unknown neurological disease among the Fore of Highland New Guinea called by them *kuru*—"trembling" or "shivering" (Vigas, 1970). *Kuru* is clinically identified by progressive ataxia of the trunk, head, and limbs and results in death within one year. The impact of the disease upon mortality rates in Fore villages has been remarkable. One hundred and sixty villages have been affected with an incidence of 5 to 10 percent of the village population. In the most severe cases *kuru* accounted for 50 percent of the deaths beyond infancy. According to Dr. Vigas, many additional deaths could be directly related to *kuru:* The second most frequent cause of death in many areas was the murder of suspected sorcerers, sorcerers whose magic was supposed to be the cause of *kuru*. The occurrence of *kuru* runs along sex and age lines. In the cases documented between 1957 and 1959, 31 percent were children under the age of 15 years and 61 percent were adult women.

Experimental studies have now shown that *kuru* is caused by an infectious virus of the brain. The virus has a long period of incubation (two years or more), can be transmitted to primates other than man, and, perhaps most important, individuals seem to have a genetic susceptibilty. The latter is based upon the observation that all known cases of *kuru* are *restricted* to the Fore and to neighbors with whom the Fore have intermarried. Dr. Vigas and D. Carlton Gajdusek, one of the earliest medical researchers into the causes of *kuru*, suggest that the virus was transmitted *via* the practice of ritual cannibalism (1957). Until the practice was stopped in 1957, the Fore have eaten their dead kinsmen during a funerary ceremony with the explanation that they wanted to remain close to their deceased loved ones. The practice was mostly restricted to close relatives of the dead. Of particular importance here is the observation that women and children consumed most of the corpse, including the internal viscera *and the brains*. Men, if they indulged at all, seldom ate the brain. Assuming that the *kuru* virus was carried dormant in the brain of the deceased, one can then argue that cannibalistic comsumption of the brain allowed the virus to be transmitted to a still living relative where it could complete the period of incubation and become active.

INTERACTION BETWEEN HUMAN POPULATIONS

The relationships among human populations are varied but include predation, competition, cooperation, and mutualism as the most important kinds. None of these should be considered as mutually exclusive categories but as interrelated. For example, it has frequently been argued that interplay between competition and cooperation is responsible for the development of civilization (e.g., Sanders and Price, 1968, pp. 94–97).

Predation

Perhaps the most unusual relationship between human populations is intergroup predation or *cannibalism*. Cannibalism is not frequent but its occurrence is fairly widespread, having been reported in several farming societies around the world and among some northern hunters and gatherers, as well as the more famous recent examples of the Donner Party in the Sierra Nevada of California and the Uruguay soccer team in the Andes of South America. In many instances the reason for cannibalism is starvation, as in the Donner Party and soccer team examples and in Eskimo cannibalism. Cannibalism as a regular source of food has also been reported among some farming societies in northern South America, particularly in Colombia, and one such group, the Picara, actually fattened up war captives before eating them. In northern South America aboriginal chiefdoms integrated ritual cannibalism into a warfare complex that included endemic warfare, human sacrifice, cannibalism, and a temple-idol cult. Warriors gained status by displaying human trophies such as heads, skulls, teeth, or even flayed and stuffed skins, and cannibalism was believed to give supernatural power to the participant. In general ritual cannibalism is based upon the concept of "soul," the sum total of an individual's abilities and capacities. By eating part of the flesh, a cannibal could take on the soul of the victim. If the victim happened to be a courageous warrior, the transfer of soul was viewed as particularly useful.

Competition

Resource-use or scrambling competition is commonplace among mobile hunters and gatherers, but little is known about its exact manifestation. Interference competition is common in sedentary groups and is better known. It is expressed in a number of ways, but sociopolitical organization and warfare are the most important. The relationship between sociopolitical organization and competition is clearly outlined by Marshall Sahlins (1961) in his discussion of the segmentary lineage system. The classic examples of the segmentary lineage society are the Tiv and the Nuer of Africa. Sahlins considers six features of the system to be distinctive: lineality, segmentation, local geneaological segmentation, segmentary sociability, complementary opposition, and structural relativity. A lineal organization, that is, a kinship group based upon demonstrated descent from a known apical ancestor, is essential to the unification, the *esprit de corps,* of local groups. Lineality is commonly associated with dependence upon resource concentrations; in this sense the segmentary lineage system is partially an adaptation to physical environment. However, only a few lineage organizations are segmentary, so we must search further for its explanation. *Segmentation* and *local genealogical segmentation* are interrelated and mean simply that the lineage is subdivided into small, localized, territorial "segments" and that adjacent segments tend to belong to the same lineage branch. For example, two adjacent Nuer villages are closer together in the lineage than two villages that are not adjacent. *Segmentary sociability* establishes relationships between the lineage segments through the "love thy neighbor" principle. The degree of friendliness and peacefulness is an inverse function of the closeness of kinship and, from what has been said above, of how close together the segments are in space. That is, close kin get along fairly well, but more distant kin are subject to more frequent squabbles. In addition the intensity of the squabble increases with distance, both kin and spatial. Violence seldom takes place among close kin but is frequent among more distant kin.

Sociability and the spatial arrangement of segments suggest that adjacent lineage segments would join together to oppose more distant segments and this is indeed what happens. Referring to this process as *complementary opposition* or the massing effect, Sahlins argues that none of the Tiv and Nuer lineages, at any level, have a real existence until segments are joined in opposition. It is at this point that the segmentary lineage system can be viewed as a mechanism of adaptation to other human populations. The Tiv and the Nuer do not consider themselves to be constrained by territorial boundaries, and traditionally they have expanded outward in all directions. Their expansion is stopped only when they encounter groups with superior military prowess. The relative ease with which both of these tribes have expanded and extended their borders at the expense of neighboring tribes is, according to Sahlins, understandable in terms of the segmentary lineage system. All that is required is a "kick" somewhere within the tribe, setting one localized segment against another in a fight for additional land. Population pressure may be sufficient or the kick may come from any other reason to gain more living space. Suppose, for example, that segment *A*, somewhere in the tribal interior, feels that more living space is needed. Then *A* enters into a land dispute with neighboring segment *B*, putting pressure on *B* to claim land from neighbor *C*. Both segments *A* and *B* join in opposition to *C*. The process continues until segments on the very border of the tribe are being opposed by a large block of interior segments joined in complementary opposition. The border segments are forced to fight with the "foreign" tribes along the Tiv or Nuer in order to regain the land taken from them by the combined interior segments. However, the principle of complementary opposition now allows the border segment to join with *all* the tribe in common opposition to the foreigner. If the foreign tribe is not able to withstand the now large military forces of the invaders, it is driven from the land and the border segments of the invaders are now able to occupy new territory, thus releasing some of the pressure from the interior. Sahlins argues that if competition from other local groups did not exist, neither would segmentary lineage. In the absence of competition local groups would split into "daughter" groups which, in turn, would simply move away into the wilderness and maintain local autonomy. This is, of course, exactly the process described for the human populations occupying the Amazon Basin. Only under conditions of social competition does an organization appear that engenders cooperation among relatively autonomous local groups—a paradox perhaps but a reasonable one.

Competition is often manifested as warfare, although the reasons for this social pathology are not well understood. In recent years many anthropologists interested in the subject have proposed population pressure as a major cause of war. Territorial conquest and increased mortality are the main reasons put forth to explain why population pressure should lead to war. Obviously, if land is in short supply, one solution is to forcefully annex the land of your neighbor. "Forcefully annex" is essential because Carneiro (1970) and others contend that human groups never give up their autonomy willingly to others. Many years ago S. F. Cook (1946) suggested that population pressure in Postclassic Mexico induced both intense warfare and human sacrifice to increase mortality and, therefore, to reduce the pressure. However, Robert McC. Netting's field work among the Kofyar farmers of northern Nigeria suggests that, although population pressure is instrumental in stimulating conflict between groups, warfare does not result directly in enough deaths to significantly reduce the pressure (Netting, 1973). Kofyar wars are betwen villages and are fought with iron-pointed throwing spears, short wooden clubs, knives, and stones. Typically, warfare takes place by conducting formal battles in border areas or in an unoccupied "no-man's land" between antagonistic villages. The battles are fought by large numbers of men who line up across the field of batle and face each other 50 to 100 or more yards apart. The hills behind the battlefield are covered with women and children who watch the gladiatorial proceedings. How lethal is Kofyar warfare? Netting argues that casualties are low; no single battle takes more than 5 lives and many take no lives at all. There is little chance that population pressure could be relieved through increased mortality. Furthermore, no territory is gained as a result of winning a battle or a war, and few individuals gain any material advantage.

Andrew P. Vayda, who has written extensively about war in non-Western societies, argues that war should be viewed as a continuous ecological process with complex, interacting causes and effects rather than as a discrete event with specific causes and effects (1974). The process of war, according to Vayda, consists of a series of phases that are graded by the type of fighting (e.g., feuding, raiding, or full fledged battles), intensity of fighting, or degree of violence. Escalation from the early stages of warfare to later stages has been seen as inevitable by some. Vayda, however, argues to the contrary; peace can be reestablished at any point in the war process. Furthermore, the reasons why a war begins are often different than the reasons why escalation occurs from one stage to another.

Cooperation

Cooperation between human populations generally takes the form of an *exchange network* for the purposes of distributing energy, materials, and information. The information function of exchange has been particularly stressed recently; exchange is viewed as social communication and the form of the exchange "conveys information to the participants about their social and spatial position" (Wilmsen, 1972, p. 1). Kent Flannery (1972a) goes further to suggest that the information characteristics of an exchange network change in an evolutionary fashion. Simple sociopolitical organizations have few information links with their social environment, irregular or "unscheduled" information flow, and few mechanisms for collecting and processing information. With increasing sociopolitical complexity, the information network becomes increasingly complex, information flow occurs at regular or "scheduled" intervals, and more and more information collecting and processing mechanisms are added. Flannery particularly stresses the importance of the evolutionary conversion of unscheduled information into scheduled information. The following example illustrates the distinction between the two.

The Tewa of the North American Southwest scheduled exchange among their pueblo sodalities, but exchange with nonpueblo groups in their environment was conducted on an unscheduled and ad hoc basis (Ford, 1972). Sodalities integrate tribal societies by cross-cutting kinship groups, and, among the Tewa, several sodalities were linked together in an exchange network based upon the needs of scheduled ceremonies. During the eighteenth and nineteenth centuries the Tewa pueblos were decimated by epidemics leaving, in many cases, insufficient sodality members to perform the traditional ceremonies. As a result, when a ceremony was scheduled by a sodality in one pueblo, members of the sodality in other Tewa pueblos were asked to help in the ceremony by providing ritual specialists. In return the host pueblo gave the assisting sodalities gifts of food and ceremonial paraphernalia. Usually, the gifts offered in exchange were needed, either economically or ritually. Consequently, the Tewa sodality exchange system not only functioned to solidify social relations but also circulated needed economic goods, thus providing a "buffer" against local economic disaster.

Tewa exchange with other tribes, their "foreign relations," was also based upon economic and ritual necessity but took place according to an irregular schedule. Trade with the mounted, nomadic Comanche is exemplary. Deer and elk meat and hides were used extensively by the Tewa but were not locally available in the eighteenth and nineteenth centuries. These goods could be obtained from the Comanche as well as buffalo heads and bison neck hair, both of which were essential to Tewa ceremonies. However, trade could not be conducted on a regular basis because of frequent hostilities with the Comanche who had a propensity to "trade today, raid tomorrow" and because the Comanche, even during peace times, were not easy to reach. They usually attended the "trade fairs" at Taos pueblo and, during these times, also visited the Tewa pueblos for trading purposes. Otherwise, groups of Tewa traders were forced to make expeditions to Comanche country on the high plains east of the Rocky Mountains. This was a dangerous venture at best because of the threat of attack from other Plains Indians and because of the constant threat of treachery by the Comanche themselves.

The importance of exchange systems in the evolution of complex sociopolitical organizations has been stressed by a number of anthropologists, including Morton Fried (1967) and Marshall Sahlins (1958). They contend that complex sociopolitical organizations arise in response to the requirements of *procuring* and *distributing* critical resources and services. William Rathje (1971) uses this theory to construct a model explaining the origin and development of lowland Classic Maya civilization. According to the model the Maya household, the basic unit of production and consumption throughout the Maya area, requires certain resources that are not found in the lowlands. The most important of these are igneous or hard stone metates, razor-sharp obsidian tools, and salt. All of these must be obtained from the highlands, but because of poor transportation and distance individual households could not engage independently in trade for these items. The households were also widely dispersed throughout the lowlands so that door-to-door selling of highland resources by traders was impractical.

Rathje argues that the most logical way of importing and distributing these resources was through the establishment of a network of local trade centers that could receive goods from the highlands and redistribute them to local households. The so-called Maya "ceremonial centers" are "nodes" in such an exchange network. Although the model thus far

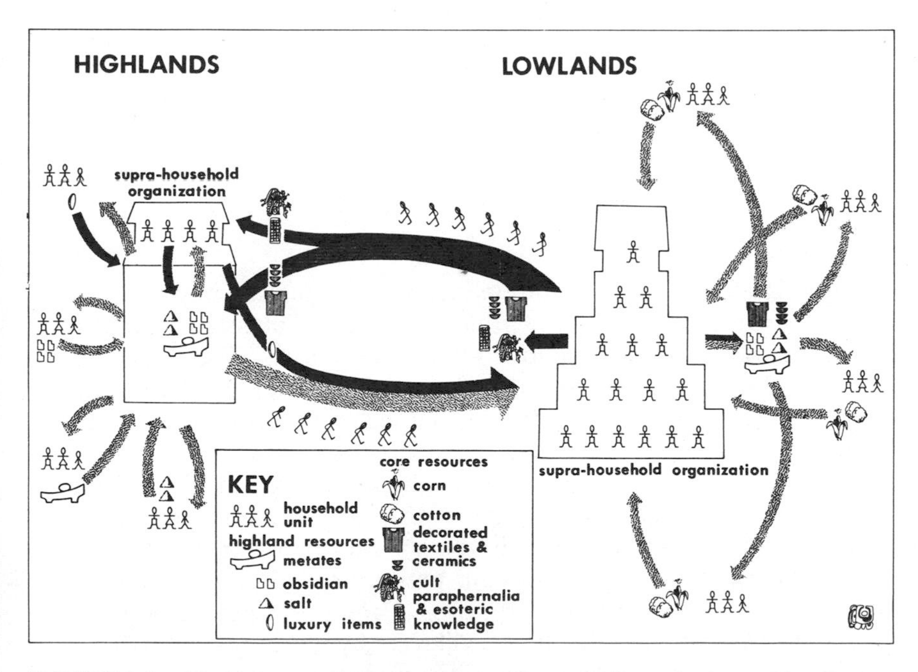

FIGURE 8.2 The Maya exchange system. (From William Rathje, "The Origin and Development of Lowland Classic Maya Civilization," *American Antiquity*, 1971. Reproduced by permission of the Society for American Archaeology from *American Antiquity* 36(3), 1971.)

explains why an exchange network should have been established in the Maya lowlands, it has not explained why these centers have an elaborate ceremonial character and why so many of the centers are situated in the heart of the lowlands in an area furthest removed from the highlands. Rathje sees a division of the lowlands into an outer or "buffer" zone adjacent to the highlands and an inner or "core" zone that is cut off from the highlands by the buffer zone. Households in the buffer zone can more easily obtain metates, obsidian, and salt through *short-distance* trade. There is no economic incentive to pass these resources on to the core zone, since both zones have the same resources. The core zone, in order to compete for highland trade with the buffer zone, must develop a system of *long-distance* trade and must create a "need" for its own manufactured products. Logistically, long-distance trade requires a specialized group or class of long-distance traders, along with the maintenance of trade routes through the buffer zone into the highlands. In itself this selects for a fairly complex sociopolitical organization. But most important was the creation of a market "need"; this was done by elaborating a ceremonial complex that manufactured luxury and ceremonial goods and gave access to the supernatural through temples altars, esoteric knowledge, and priests. If Rathje's model is correct, the lowland Maya were among the first great marketing psychologists. Figure 8.2 illustrates the exchange system.

Mutualism

The classic example of mutualism among human populations is Fredrik Barth's study (1956) of ethnic relations in Pakistan. Two of the groups that Barth studied are the Pathans and Gujars. The Pathans are advanced plow farmers who occupy the floors of high mountain valleys. They are organized into centralized, unilineal, and multicaste societies, which include specialized occupational subgroups that trade their services for payment in food and materials. In addition political clients and economic serfs are tied to the Pathans. One of the specialized occupational subgroups is the Gujars. The Gujars are tribal herders of sheep, goats, cattle, and water buffalo who practice transhumance and true nomadism within Pathan territory. Most Pathan villages have a small group of Gujars who are tied to the villagers by a client or serf relationship. The Gujars care for animals in the village and supply the villagers with milk products and meat for consumption and manure for farming. Farming labor is also supplied by the Gujars, particularly during the critical period of transplantation of wet rice. In exchange for these services and goods, the Pathans give grain, textiles, and hardware but, most important, they give access to the pasturelands underlying pastoral nomadism.

Barth argues that cooperation of this sort arises to prevent competition according to the principal of competitive exclusion. That is, the Pathans and Gujars coexist in the same area because they have different but complementary life styles. However, others have argued that symbiotic relations between farmers and herders are based, not on mutual agreement, but on the political balance of power. In other words, cooperation is an adaptation to social environment. Data given by Daniel Bates (1971) for the Sacikara Yoruk of southeastern Turkey support the argument. The Yoruk are sheep herders who follow a pattern of transhumance similar to the Iranian Basseri described by Barth. Yoruk migrations are closely integrated into the crop cycles of village farmers occupying the pastures used for grazing, moving their herds into each seasonal pasture only when there is little danger of the sheep's damaging the crops. Furthermore, the pasturelands are owned by the farmers and rented annually to the Yoruk for cash. Bates shows that this patten is not based upon mutual adjustments and settling into different niches but is based upon the political realities of the outside world. Since the middle of the nineteenth century, the Turkish state has conducted a program to force sedentism on the once large and powerful Kurdish and Turkmen nomads occupying the region, and the program is nearly completed. The Yoruk moved into the region after the program had started and were allowed to remain nomads and to use the pastures and migration routes by the state because they presented no political threat, as had the Kurdish and Turkmen tribes. However, Yoruk nomadism is highly constrained by political exigencies. The state granted ownership of the pastureland to the village farmers, and the Yoruk were forced to pay rent to these landlords. Without state backing, the farmers would not be able to enforce the rent and a new kind of adjustment would have to bẹ worked out. The importance of balance of power is also exemplified by the fact that the Yoruk were required to pay damages, in cash, to the village farmers if their herds destroyed crops. As a result the Yoruk must schedule their movements from summer to winter pastures in such a way as to minimize crop damage, even though the schedule does not coincide with optimal grazing times. Removing the state government from the social environment would undoubtedly lead to a change in migratory schedule and to a contraction of farming land away from those areas most likely to be damaged by the herds.

SUMMARY

Human population ecology is not only the study of how humans get along in their physical environment but also how they interact with other ecological populations. The type of interaction may be either negative, positive, or neutral, and it may involve either human or nonhuman populations. Predator-prey relationships are the most common between humans and nonhumans, including hunting wild animals or gathering wild plants on the part of humans and predation by microorganisms on humans. Nevertheless, there are some clear examples of mutually beneficial relationships. The most important relationships among different human populations are predation, competition, cooperation, and mutualism. Competition is particularly common in relationship to scarce resources and is often expressed as warfare. At the same time the cooperation expressed in, and reinforced by, exchange networks and mutualistic relationships are equally important as ways of coping with other human populations.

NINE POPULATION GROWTH

In recent years *population growth* has been used more and more often by ecological anthropologists to explain why social and cultural changes take place. On the one hand, the changes are viewed as self-regulating, to keep population size within the limits of available resources. On the other hand, the changes are interpreted as positive feedback processes operating to increase the environment's carrying capacity and thus to accommodate a growing population. The origins of farming and the state have been explained in this way, for example, and many anthropologists feel that the explanation of evolutionary changes of this kind is the main contribution of the ecological approach. In this chapter the immediate demographic causes of population growth are discussed. In general a population grows, stays the same, or declines because of variation in the birth rate, the death rate, and the rate of migration into and out of the population. Each of these rates has innumerable biological, social, and cultural causes, some of which are due to ecological relationships. The following two chapters will take up population growth as an ecological process.

FERTILITY RATES

The rate at which new members are being added to a population through birth is commonly referred to by demographers as the fertility rate. (Note, however, that some demographers and biologists use *fecundity rate* to refer to the actual reproductive performance of a population and use fertility rate to mean the potential capacity of the population to reproduce. Be sure that you understand in which sense the term is being used.) Unfortunately, fertility rates are measured in a vast number of ways. The *crude fertility rate* is the easiest to calculate but is also the least accurate; it is simply the number of births for each 1000 members of the population for a single calender year.

Crude fertility rates do not take into consideration different rates of reproduction due to age and sex. However, a number of *specific fertility rates* do make these corrections. *Age-specific fertility rates* are calculated from the number of births for each 1000 members of the *same* age group during a calendar year. Several specific rates are commonly used. The *general fertility rate* is the number of annual births per 1000 *women of child-bearing age* (usually ages 15 to 49). The *total fertility rate* is the actual reproduction of a group of women to whom a set of age-specific fertility rates apply. Conventionally, the group is taken to be a *cohort* of women (all born within the same time period, usually a calendar year) passing through their entire reproductive span. If the age-specific birth rates are available for *each* year of the reproductive span and if the rates are given as births per 1000 women, then the total fertility rate is simply the *sum* of the age-specific rates for all the years of the reproductive span. However, age-specific rates are usually given for a *five*-year period so that the procedure is somewhat different. Since each woman is assumed to reproduce at the same rate for *each* of the five years, her actual reproduction during this time is the fertility rate for a given five-year period multiplied by five (the number of years in the age group to which the fertility rate applies).

The *gross reproductive rate* is the number of *female* children born to each 1000 females during their reproductive span. In effect the gross reproductive rate is the total fertility rate corrected for sex ratio. Thus in a large human population there will be 100 females out of every 205 births or a proportion of .488. Multiplying the total fertility rate by .488 gives the gross reproductive rate. However, any anthropologist interested in demography knows that a sex ratio of 105 seldom holds for small populations. For this reason care must be taken to use the actual sex ratio when total fertility rate is converted to gross reproductive rate.

Finally, the *net reproductive rate* (NRR) is the number of female offspring that actually replaces each female of the previous generation. Thus a NRR of 1.0 means that the population is just replacing itself by adding 1 female each generation to replace each one lost, assuming that the age structure is stationary. An NRR of 2.0 means that the population is doubling each generation, and so forth. The procedure used is to multiply the fertility rate for each age category (age-specific fertility rates) by the probability of surviving long enough to enter that age category. Then the products for all age categories are summed together and that total is multiplied by the proportion of total births that are female.

Fertility rates vary greatly from one human population to another and from one time period to another. We have little reason to believe that human populations differ significantly in fecundity (i.e., the intrinsic biological ability to reproduce), although individuals may. However, idiosyncratic, social, and cultural behavior have quite an impact. The maximum rate of fertility in well-known human populations is about 10.4 births per woman, a total fertility rate documented for the Hutterites (Eaton and Mayer, 1953). A rate this high is unusual, at least in Western populations, and is reduced either by (1) shortening the number of years that women are exposed to the "risk" of pregnancy, or by (2) reducing the chances of conception during the time that they are exposed (Howell, 1973, p. 254). Exposure to the possibility of pregnancy is mainly affected by how early and how often marriage takes place (most pregnancies still occur within marriage, although there is some obvious error here), how often and how soon after marriage that divorce or widowhood occurs, and how often and how soon widows and divorcees remarry. Answers to all of these questions help determine the actual period of time during their reproductive years that women are married and thereby exposed to the possibilty of pregnancy. Reducing the chances of conception during the married years is another way of decreasing fertility rates. Some of the relevant practices and conditions follow (Benedict, 1972; Newman, 1972):

1. *Post-partum taboo.* Sexual intercourse is not sanctioned until after the mother has stopped lactation or until a specific period of time since birth has passed.

2. *Ceremonial abstinence.* Sexual intercourse is prohibited during ceremonial occasions, for example, just before the hunt. Also included would be abstinence for philosophical reasons, such as the belief that intercourse reduced the power or strength of an individual or offspring, or for "religious" reasons.

3. *De facto abstinence.* Sexual intercourse takes place, but sperm is not transferred due to the practice of *coitus interruptus* (withdrawal), *coitus obstructus* (manual obstruction of the semen duct), and *coitus reservatus* ("self-control" method in which ejaculation is prevented by individual will).

4. *Rhythm.* Sexual intercourse is not practiced during periods when the woman is assumed to be fertile.

5. *Contraception.* Sexual intercourse takes place but mechanical methods are used to prevent conception. Methods include the use of the male sheath, female pessaries (obstructive plugs or caps placed over the cervix or entrance to the womb), and the female douche (vaginal cleansing with an acidic substance).

6. *Age.* The probability of conception during sexual intercourse varies with the age of the participants.

7. *Health.* The probability of conception also varies with the general health of the participants, particularly their state of health as determined by nutrition and by psychological stress (e.g., psychological inability to have an erection).

8. *Polygamy.* Polygamous marriages reduce fertility rates because of lower frequency of coitus and older ages of husbands. No known effect comes from polyandry.

9. *Abortion.* Intentional killing of prenatal offspring. The most common methods used are chemical (use of drugs, herbs, etc.), mechanical (disturbing the uterine contents with mechanical devices), "activity" (jumping, rolling, running, etc.), and magical.

The social matrix in which the causes of fertility may be found is usually so intricate that it defies easy understanding. In rural Ireland, for example, it has been shown that an exceedingly low birth rate is caused mostly by late marriage and widespread celibacy (Arensberg and Kimball, 1968; Messenger, 1969). The social conditions contributing to both are complex and include a system of impartible inheritance (in which farming land is given only to one son and not until the father is dead, is willing to, or is forced to pass on the land), male solidarity, jealous and domineering mothers reluctant to accept a daughter-in-law into the household, and sexual puritanism. According to John Messenger the last is the most characteristic personality trait of Irish males and has its origin in historical (such as the influence of ascetic monasticism and Jansenism), sociocultural (such as the Oedipus complex and male solidarity), and psychological (such as masochism) variables (1969, p. 107). The reluctance to engage in sexual activities not only brings about late marriages but also reduces the chance of conception after marriage takes place.

Fertility rates can also be affected by diet. Experimental studies, for example, have shown that malnutrition caused by a deficiency in calories, protein, vitamin B, and specific minerals such as iodine can produce hypofunctional gonads in both sexes and thereby reduce fertility rates (Katz, 1972, pp. 357–58). In addition there is a suggestion that the age of menarche (the onset of puberty) is later in persons with an inadequate diet, thus shortening the total reproductive period. A change in diet can lead to a reduction in fertility rates by shortening the interval between births, according to recent studies of the Kung Bushmen (Lee 1972b, pp. 340–42). The relationship hinges upon the length of the breast feeding period. Breast feeding suppresses ovulation so that the length of the lactation period is important in determining fertility rates. !Kung populations with a traditional subsistence pattern have a very long breast feeding period because the wild foods that are gathered are too hard for children to digest until they reach the age of three or four. Consequently, survival during early childhood requires breast feeding. On the other hand, the incorporation of softer foods into the diet, particularly milk from domesticated animals, allows the infant to be weaned earlier. For this reason the *required* breast-feeding period is considerably shorter, more frequent births are possible, and the interval between births tends to be reduced. The resultant higher fertility rates cause population increase even though group mobility and available food remain the same. This suggests that digestibility differences in wild foods may be an important clue to differential rates of fertility among hunting and gathering populations.

MORTALITY RATES

The rate at which members of a population are being removed through death is the mortality rate. As with fertility both crude and specific mortality rates are used to measure the death characteristics of a population. The *crude mortality rate* is the number of deaths per 1000 individuals for each calender year and is calculated in exactly the same way as crude fertility rates. A *specific mortality rate* is the number of deaths within a population defined by age and/or sex during a given period of time. For example, the number of annual deaths per 1000 individuals in the 20 to 24-year age group is an age-specific mortality rate. Five-year intervals are the most commonly used in calculating mortality rates but an exception is the *infant mortality rate,* the number of deaths per 1000 individuals under one year of age.

Mortality rates also vary widely from one human population to another and, as with fertility rates, are often associated with social and cultural practices, including the following:

1. Infanticide. The offspring is intentionally killed immediately or soon after birth.

2. Senilicide. The old are intentionally killed or commit suicide.

3. Warfare. Usually increases mortality among young adult males.

4. Ritual. Human sacrifices and cannibalism in a ritual context.

5. Magical. Death due to psychological loss of the will to live.

6. Invalidicide. Invalids are intentionally killed.

As an illustration of the specific causes of human mortality, mortality statistics for two Yanomamo (Venezuela and Brazil) villages are listed in Table 9.1. The impact of social conditions on mortality patterns is suggested by a recent demographic study of the town of Agnone in southern Italy (Douglass, n. d.). One parish census, taken between 1841 and 1863, gives data on the number of infants who died before their second birthday. If the parish is considered as a whole, one out of every 3.31 infants

TABLE 9.1 CAUSES OF DEATH AMONG THE YANOMAMO

	Shamatari Village		Namowei-teri Village	
Informant Explanation	Males	Females	Males	Females
"Natural" causes	0	0	0	2
Epidemics	32	78	72	81
Dysentery	1	1	14	5
Warfare	52	5	44	9
Duels	1	1	5	0
By husband	0	0	0	1
Snake bite	1	3	4	2
In childbirth	0	0	0	1
Respiratory infection	1	5	3	0
Hayaheri (abdominal pain)	0	1	6	1
Old age	7	4	8	3
Sorcery	27	11	15	5
Crushed by falling tree	1	0	1	0
Hekura (spirits)	6	5	13	20
Measles	0	1	0	0

Source. Napoleon A. Chagnon, *Studying the Yanomamo,* New York: Holt, Rinehart and Winston, 1974.

did not survive this period; however, that average does not hold for different social groups. The infants of professionals and merchants had a greater chance of survival, only one out of every 3.78 dying. A similar deathrate was found among the children of artisans (one death for each 3.89 births). By contrast the *contadini* (peasant) infants had a much lower chance of surviving past their second year; one out of each 2.79 born died. The reason seems to be a more unsanitary life style. Thus the Italian physician Carlo Barbieri, writing in the early part of the nineteenth century, notes that the peasant "always shares his room with the chickens, and his bed is over the pig sty and donkey stall, for this reason throughout the night he breathes an unsanitary air" (quoted in Douglass, n.d.).

A similar difference prevails for the life expectancy of adults. In another parish census from Agnone, taken between 1820 and 1829, data show that peasants and manual laborers survived to a mean age of 48.78, artisans to 43.55, and the *galantuomini* (upper class) to 60.11. The relatively long life of the *galantuomini* is apparently due to "superior diet, better medical care and freedom from hard physical toil" (Douglass, n.d.). In contrast, the short adult life of the artisans is due to their poor working conditions. Working hours were long and the workshops provided an unhealthy atmosphere because of poor ventilation and extreme temperatures. The poor conditions not only directly caused illness and death but also indirectly, by encouraging excessive drinking.

The probability of *survivorship* from one age group to the next is a measure of relative mortality within a population that is quite useful for prediction. Let us assume that we are tracing the life history of a cohort of 1000 through successive age intervals of five years. At the end of the first five years we count the survivors and observe that 900 remain to pass into the next five-year interval. Because 100 have died, the *probability* of a person in this interval dying during the first five years is 100/1000 or .100. Conversely, the probability of surviving is 1 − the probability of dying = 1 − .100 = .900. After another five years another count is taken of the cohort, and we observe that only 700 have survived, indicating that 900 − 700 or 200 persons have died during this period. The probability of dying in the second five-year interval is therefore 200/900 (remember that only 900 entered this age interval) or .222 and the probability of surviving is 1 − .222 = .778. The procedure is continued in successive age intervals until all members of the cohort have died.

Life tables are conventionally used to describe the survivorship characterisics of a population. Table 9.2 illustrates how a life table is arranged. Five symbols are used to designate the columns: (1) x is the age in years, (2) lx the number of persons surviving to age x, (3) dx the number

TABLE 9.2 LIFE TABLES FOR VIRGINIA CITY, NEV., IN 1880 [a]

Age x x	Survivors to Age x lx	Deaths Between x and $x+1$ dx	Probability of Death Between x and $x+1$ qx	Probability of Survival Between x and $x+1$ px
Males				
0	177	41	.232	.768
5	136	2	.015	.985
10	134	7	.052	.948
20	127	11	.087	.913
30	116	34	.293	.707
40	82	45	.549	.451
50	37	26	.703	.297
60	11	9	.818	.182
70	2	2	1.000	.000
Females				
0	95	41	.432	.568
5	54	2	.037	.963
10	52	6	.115	.885
20	46	8	.174	.826
30	38	15	.395	.605
40	23	11	.478	.522
50	12	7	.583	.417
60	5	0	.000	1.000
70	5	5	1.000	.000

[a] *Data from Lord, 1883.*

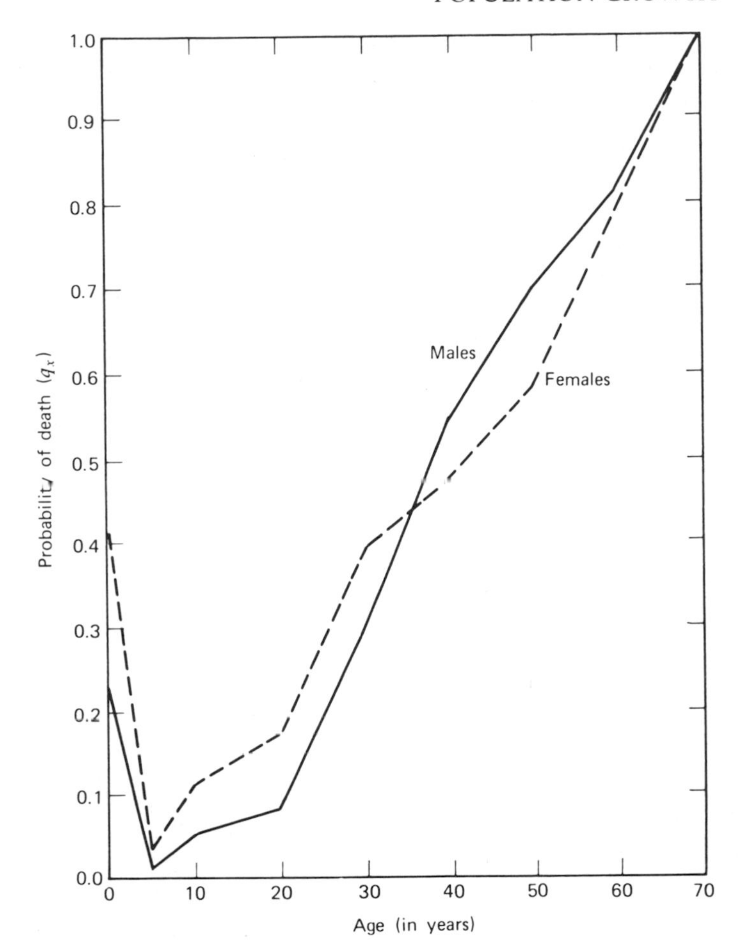

FIGURE 9.1 J-shaped mortality curves for Virginia City, Nev., in 1880. (Data from Lord, 1883.)

of deaths between ages x and $x + 1$ (the next successive age), (4) qx is (dx/lx) the probability of dying at age x, and (5) px is ($1 - qx$) the probability of surviving at age x.

Several kinds of curves can be constructed from the data in life tables. The most common are *mortality curves* and *survivorship curves*, (R. Smith, 1974, p. 308). Mortality curves are simple plots of the age-specific probabilities of dying (qx) against age. Since in higher animals the death rates of younger age groups is low in comparison to older-age classes, the curve is typically J-shaped. (See Figure 9.1) The mortality curve is particularly useful when there is some doubt about the accuracy of demographic statistics on some age groups, since qx is calculated strictly from the number of individuals who entered each age group and died there; it does not accumulate previous errors. Consequently, curve-fitting procedures can be used to approximate mortality characteristics of a population even if some of the data are inaccurate. Both anthropologists and wildlife

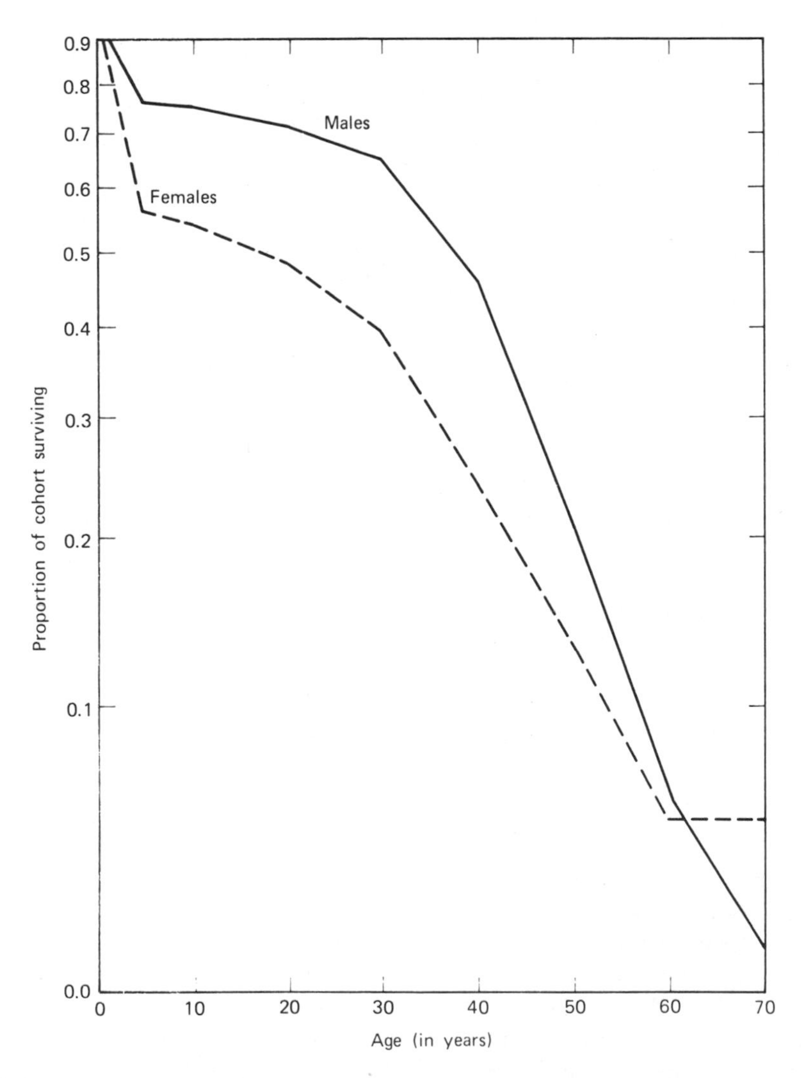

FIGURE 9.2 Survivorship curves for Virginia City, Nev., in 1880. (Data from Lord, 1883.)

biologists, for example, have difficulty in estimating death rates for the very young age groups, the former often because of reluctance on the part of informants to mention infant deaths.

Survivorship curves plot the number of survivors in a specific age group (lx) against age. The vertical axis, which may be scaled either arithmetically or logarithmically, is used to plot lx while the horizontal axis, scaled arithmetically, is used to plot the age classes. If a logarithmic vertical scale is used, a survivorship curve represented by a diagonal line means that the mortality rates are the same for all ages. On the same scale a convex survivorship curve shows that most members of the population reach maximum life expectancy before dying and a concave curve shows that most members die early in life. Figure 9.2 illustrates the survivorship curve. Again, however, remember that the survivorship curve must be used with caution. Unlike mortality curves those of survivorship accumulate errors, whether due to sampling or other problems. The small size of many anthropological populations, skeletal or living, subjects them to large sampling errors, particularly in some age classes so that the survivorship curve may not be overly accurate (Moore, Swedlund, and Armelagos, 1975).

RATES OF POPULATION GROWTH

If fertility rates do not exactly match mortality rates, other things being equal (migration rate, age structure), the size of a population will either increase or decrease. The rate of growth or decline can be expressed in several ways, but the *rate of natural increase* is one of the most commonly used. The rate of natural increase is simply the difference between the crude fertility rate and the crude mortality rate converted to a percentage. For example, if the crude fertility rate is 20 per 1000 and the crude mortality rate is 10 per 1000, the rate of natural increase is $20 - 10 = 10$ per 1000. Dividing by 10 to give the amount of change (increase or decrease) per 100 individuals, we find that the correct rate is 1 percent. If the fertility rate is greater than the mortality rate, the rate of natural increase is positive; that is, the population is growing. On the other hand, if the fertility rate is less than the mortality rate, the rate of natural increase is negative and the population is declining. (See Table 9.3) The rate of natural increase is more easily understood if it is viewed as the number of years it takes to double population size. Although not extremely accurate the rate of natural increase can be converted to "doubling rate" by dividing it into 70. Thus a natural increase rate of 2.0 means that the population size will double every 70/2 or 35 years.

TABLE 9.3 RATES OF NATURAL INCREASE IN SEVERAL MODERN HUMAN POPULATIONS[a]

	Natural Increase
Colombia	3.4
Venezuela	3.4
Rhodesia	3.4
Morocco	3.4
Algeria	3.3
Pakistan	3.3
Mexico	3.3
Kenya	3.0
Ghana	2.9
Costa Rica	2.7
Guatemala	2.6
Uganda	2.6
India	2.5
Hong Kong	2.4
Ivory Coast	2.4
Guinea	2.0
Singapore	2.2
Sri Lanka	2.2
Taiwan	2.2
Egypt	2.1
Malagasy Republic	2.1
Upper Volta	2.0
Australia	1.9
China (Peoples Republic)	1.7
Chile	1.7
Japan	1.2
United States	0.8
Italy	0.7
France	0.6
United Kingdom	0.3
Sweden	0.3
Belgium	0.2
Luxembourg	0.0

[a] *Data from Boughey, 1975: 277, Table 7-11.*

DEMOGRAPHIC RATES AND POPULATION STRUCTURE

Ferility and mortality rates affect the age structure of a population. Changes in mortality rates have little effect upon age distribution as long as the changes are reasonably gradual. In Western populations it has been documented that gradual and constant changes in mortality rates in the *reproductive* age classes have almost no effect on age structure (Pressat, 1972, pp. 298–301). However, a constant decrease in mortality rates in age classes *below* the reproductive span tends to lower the age distribution and to make the age pyramid more bottom-heavy. Conversely, a constant decrease in mortality rates in age classes *above* the reproductive span raises the age distribution and produces an age pyramid that is more top-heavy. In Western populations demographers have observed that changes in mortality rates in age classes below and above the reproductive span have just about balanced each other so that no lasting effect on age distribution could be documented.

Fertility is much more influential on age pyramids. Many demographic studies have documented that decreasing fertility rates (which obviously must take place in the reproductive age classes) raises the age distribution and is probably responsible for the increasingly top-heavy age pyramids of Western populations during the first half of this century (Pressat 1972, pp. 298–301). Increasing fertility rates have an opposite effect. The biostatistician A. J. Lotka (1925) demonstrated that if age-specific birth- and deathrates in a population are *constant,* the proportion of the population in each age class will eventually stabilize and remain absolutely constant thereafter. Such populations are said to be *static.* A *relatively* constant age structure is also reached if birth and death rates are increasing or decreasing but are doing so systematically. In this case the exact proportions of each age class in the population do not remain the same from generation to generation, but the proportion of each age class to the others remains constant. *Stable* populations such as these have been studied in some detail and *Model Life-Tables* have been constructed for them giving the age composition to be expected for different birthrates. (It has been shown that even if deathrates vary, there is little effect upon the age structure.)

DEMOGRAPHIC RATES IN SMALL, NONLITERATE POPULATIONS

Anthropologists are most likely to be working with small populations that do not have written records. In some cases the population will be living; in other cases the population is represented only by skeletal remains. While the demographic problems encountered when working with each kind of population are unique in some ways, sampling errors and errors of description are likely to be encountered in both.

The real difficulty with small populations is that the conventional demographic rates have limited significance for predicting the population's future. Crude birth- and deathrates, which are useful indicators of population dynamics in large populations, vary so much from year to year in small populations that they may have little meaning. Further, age-specific rates and the indices based upon them (e.g., total fertility rate and gross reproductive rate) fluctuate even more wildly than the crude rates. Demographers recognize that migration and age composition are probably the most important determinants of vital rates in small populations. For this reason it is essential that the anthropological demographer be familiar with the impact of age composition upon vital rates. Pressat (1972, p. 350) gives some suggestions. Since mortality rates in the oldest age classes are generally the highest, a small population with a top-heavy age pyramid will have higher crude mortality rates than similar populations with bottom-heavy age pyramids. Consequently, the *percentage* of the population in the oldest age classes must be considered when explaining mortality. On the other hand, cultural factors can cause very high mortality rates in other age classes, if so, the percentage of the population in these age classes will have a significant impact upon overall mortality rates. An example would be populations with endemic warfare or enduring a period of intense warfare, which would result in a high mortality rate among young adult males.

The percentage of the population in the youngest age classes is not an accurate measure of fertility in small populations. A more accurate measure is the *fertility ratio* or the *child-woman ratio* (Eaton and Mayer, 1953). The total number of individuals in the youngest age class (e.g., under five) is divided by the total number of women in the reproductive age span (e.g., 15 to 49) and multiplied by 100. In this way fertility due both to legitimate and illegitimate births is measured and a large proportion of women past the reproductive stage will not be given too much weight in the calculation of fertility rates.

Even though detailed information about the age composition of small populations is essential, anthropologists working in nonliterate societies often find this information hard to collect. Kenneth Weiss (1973) has made an extensive study of age errors likely to be made in such populations and offers several suggestions as to how they might be mitigated. Fertility data are likely to be inaccurate because infant death and death in younger age classes are not always reported. The reasons are several but often revolve around the fact that in many societies children are not considered to be "people" until they reach the age of two or three years. Consequently, answers to questions about the number of births will frequently include no mention of infants who have died at birth, who have died early, or who have been killed intentionally (infanticide). This type of error will lead to a top-heavy age pyramid and is an explanation for this distribution that must be carefully considered. A second source of error, according to Weiss, is temporary migration of groups of individuals out of the population. Group absences are particularly common in cash labor situations where young adults move to cities or large plantations and do not return for many years. If the census is taken during a short period of time, another common "missing" group is males who are on an extended hunting trip. Age classes making large indentations in an age pyramid may be better explained by temporary migration than by the birth of an unusually small cohort. If there is supporting evidence, the missing segment may be approximated by using conventional curve-fitting methods.

Small populations are particularly vulnerable to *stochastic* (random) fluctuations in vital rates. That is, systematic factors affecting births, deaths, and migrations are often overridden by accidents that in large populations would not have a significant affect. Since the oldest age classes are generally quite small, accidental death is likely to be an important factor affecting age-specific mortality rates. For this reason Weiss recommends that individuals over 55 should be ignored in anthropological studies of small populations. However, accidents in other age classes can also cause a dramatic reduction, and the field worker must be constantly on the alert for such events.

Weiss feels that the greatest problem faced by anthropological demographers is the inability of Westerners to correctly age individuals in nonliterate societies by visual means. Field workers have generally been completely unable to accurately age individuals over 30 years of age in many non-Western populations with the result that the older age classes are often grossly underrepresented. As a result some recent demographic surveys of the Bushmen of the Kalahari Desert in South Africa have grouped together *everybody* who is judged to be over 30. A second problem with estimating age shows up in the number of fertile adult females. There is a definite tendency to place all members of this group into the 20-to-30-year age classes, thus underrepresenting females in their teens and in their forties (some of whom will also be reproducing). Weiss suggests drawing a smooth curve between the young and old age classes if a female peak in the 20 to 35 age classes is evident. Inaccurate aging is probably a better explanation than an abnormally large cohort. Finally, anthropological demographers have a definite preference for ages ending in 0 or 5, (e.g., 10, 15, 20, 25, etc.). Consequently, if age intervals of less than 5 years are being used, age classes ending in 0 and 5 will contain more individuals than those ending in other numbers. This problem can be avoided by grouping all age classes into five-year intervals that are centered at the preferred ages, that is, those ending with 0 or 5. For example, if 15 is a preferred age, the appropriate five-year class would be 13 to 17.

SUMMARY

Population growth can be understood as the product of birth, death, and migration. The rate at which new members are added to a population by birth is given by several demographic measures, including the crude fertility rate and several specific fertility rates. Fertility rates are caused by human biology and by a wide range of social and cultural factors such as the age of marriage, celibacy, contraception methods, ritual, and diet. The removal of individuals from a population by death is measured by mortality rates and is also affected by a variety of biological, social, and cultural factors. Mortality rates differ dramatically from one age group to another, and life tables are often constructed to show how populations differ in the age that individuals die. Ideally, life tables reflect the life styles and biology that cause different patterns of survival and, for this reason, have been used by insurance companies to set insurance costs for each age group. Fertility and mortality rates are closely related to the age structure of a population. If both remain constant or change at the same rates, the age composition of the population will eventually become stable. Along with the migration rate, the rate of natural increase measures how fast the size of a population is changing over time.

Demographic rates are difficult to measure in the small, nonliterate, and sometimes nonliving populations in which anthropologists are likely to be working. Small populations are subject to random fluctuations in birth and death rates that may drastically change membership in age and sex classes. Written records are usually not available and age determination can be difficult indeed. Membership in younger age classes is particularly difficult to ascertain because of unreported infanticide or death at birth and poor preservation of skeletal material (in skeletal populations). Furthermore, members of some age and sex classes may be absent at the time the anthropologist takes a census, because of outside wage work, hunting, or other activities. Various methods of correction are necessary to give a good demographic approximation of such populations.

TEN
GROWTH AND REGULATION

The demographic rates of human populations are not constant but change over time in such a way that a few basic *patterns of growth* are described. One common pattern is the J-shaped curve or the "boom-bust" curve. The population grows rapidly at first, approximating an "exponential" growth pattern, but then crashes to a much smaller size on the heels of rapid negative growth. Another common growth pattern is the S-shaped, sigmoid, or logistic curve. In this case the population grows slowly at first, then goes through a period of rapid exponential growth and finally levels off at a size that can be permanently supported by the environment, that is, the carrying capacity. The growth pattern resembles an "S," thus the name. Figure 10.1 illustrates the two patterns of growth, along with some variants.

DEMOGRAPHIC TRANSITION THEORY

Classic demographic theory has presented a picture of human population growth with a slow, gradual rise from the beginnings of man until about 1750 when the advent of the Industrial Revolution stimulated a population "explosion." However, during the twentieth century the growth rate in industrialized Western countries has leveled off. The pattern of growth is S-shaped. Changes in the growth rate from preindustrial to fully industrialized societies is described by the three stages of the so-called theory of the *demographic transition*. Stage One ("pretransitional") is marked by zero population growth and is associated with preindustrialized societies. Fertility rates are very high but are offset by very high rates of mortality, caused principally by losses to uncontrolled diseases. Both fertility and mortality rates are reduced in Stage Two ("transition") because of the introduction of health care services and birth

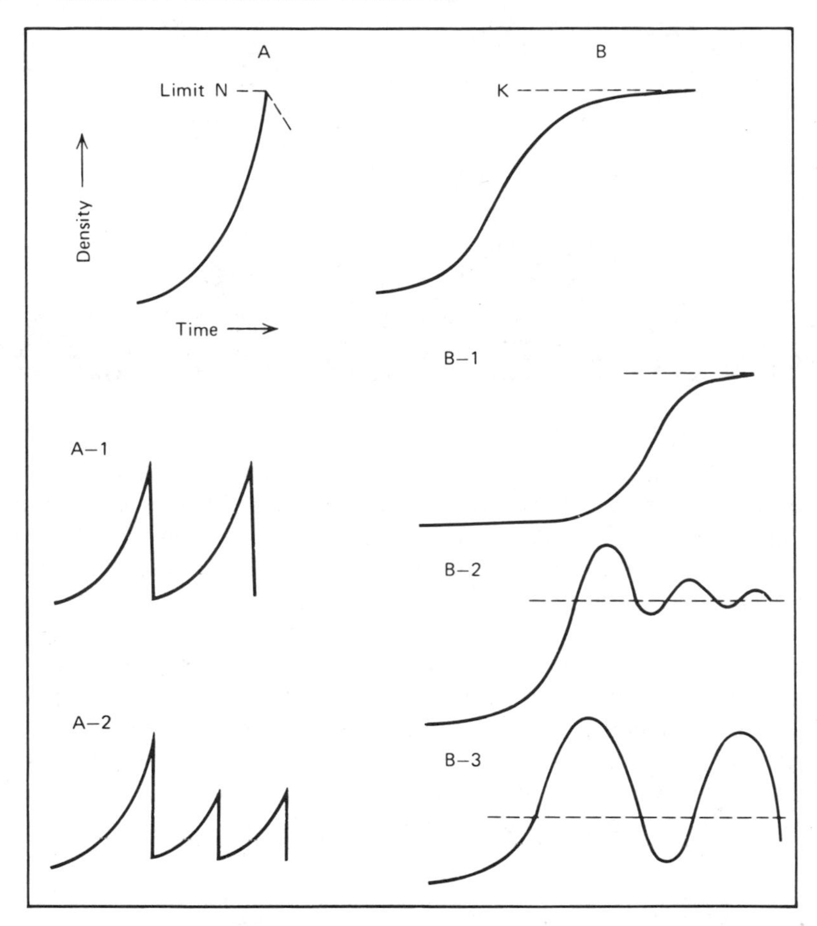

FIGURE 10.1 Basic patterns of population growth. (a) J-shaped or "boom-bust" growth with variants A-1 and A-2. (b) S-shaped or logistic growth with variants B-1, B-2, and B-3. (From Eugene Odum, *Fundamentals of Ecology,* 3rd ed., Philadelphia: W. B. Saunders Company, 1971. Copyright © 1971 by W. B. Saunders Company. Reprinted by permission of the author and publisher.)

control devices and education. However, mortality is reduced much faster than fertility and the population surges upward during most of this stage.

The transition stage is divided into three substages: early, middle, and late. Rapid growth occurs during early transition because mortality rates have been drastically reduced by the introduction of medical service, but fertility rates remain the same as in the pretransitional stage. The subsequent decline in mortality is mainly due to the much greater chance of infants' surviving into the first two or three years of life. Instead of 70 or 75 percent surviving, 96 or 97 percent survive. Fertility rates begin to decline with the introduction of birth control devices and information, and the population enters the middle transition stage. Growth is slowed down but is still high. Finally, during the late transition stage mortality rates have been reduced to a low level and fertility rates are only slightly higher. Population size is beginning to level off. The proposed reasons for the drop in fertility are directly connected with industrialization and mechanization. In preindustrial societies children are potential producers, helping on the farm or assisting in the hunt. However, with the advent of industrialization children were no longer producers but consumers upon whom capital had to be expended. Small families were valued over large families. Even in rural areas the introduction of machines driven by domestic animals or fossil fuels essentially eliminated the need for large families. Consequently, advantage will be taken of birth control methods because of a change in values induced by a technological revolution. The final stage of the demographic transition ("posttransition") is the culmination of the changes taking place in Stage Two. Both birth and death rates have been reduced to a very low level, and population growth is once again zero.

Some recent writers have contended that the three-stage demographic transition does not adequately describe what has happened during the industrialization of twentieth-century America (e.g., Easterlin, 1966; R. Hall, 1972). Population stability has not yet been reached, contrary to expectations, and fertility rates appear to fluctuate with the condition of the economy. Thus during the 1960s the crude fertility rate fell dramatically (to a low of 17.8 in 1967) as economic conditions in the United States worsened. The economist Richard Easterlin (1966) attributes most of the decline in fertility to the white urban middle class. The anthropologist Roberta Hall (1972) argues that another reason for the reduction in fertility is the increasing number of women who joined the labor force in recent years. Considerations of the amount of time necessary to bear and raise children would instigate an antinatalist attitude within this group. Hall goes on to propose Stage Four and Stage Five to the demographic transition. Stage Four is presently being experienced within American society and is marked by a generally low but constantly fluctuating birthrate. The birthrate is basically controlled by economic conditions and is occasionally high enough to produce a population "explosion." Stage Five is hypothetical and describes the achievement of population stability. Birthrates are low and stable because of technological automation, the emergence of the *individual* as a basic economic unit (replacing the nuclear family or lineage), and increased participation of women in the working force. However, in contrast to many geneticists Hall believes that differences in individual fertility will still be large and susceptible to natural selection, the main difference being between women who work outside the home and those who do not.

Reflecting the views of several recent anthropologists concerned with demography, Steven Polgar (1972) questions the following basic tenets of the demographic transition: (1) Since the advent of man, there has been a slow gradual increase in population size with a single population explosion caused by the Industrial Revolution; and (2) both birthrates and deathrates were very high until the beginning of industrialization. The first tenet is widely disregarded by prehistorians. Archaeological evidence suggests that at least two other population surges occurred during man's life on earth, one associated with the invention of tools, the other with the invention of farming (Figure 10.2). In addition there is also evidence that population explosions were closely associated with the advent of *intensified* food collecting, such as presumably occurred during the Mesolithic, and with early urbanization (e.g., Braidwood and Reed, 1957). The most significant points to be made about human demography prior to the Industrial Revolution have been summarized as follows: (1) The first *major* population surge occurred with increasing *sedentism* at the beginning of the Neolithic (associated with intensified food collecting first and then farming); (2) population growth in the past had important cultural and biological consequences, particularly in the intensification of

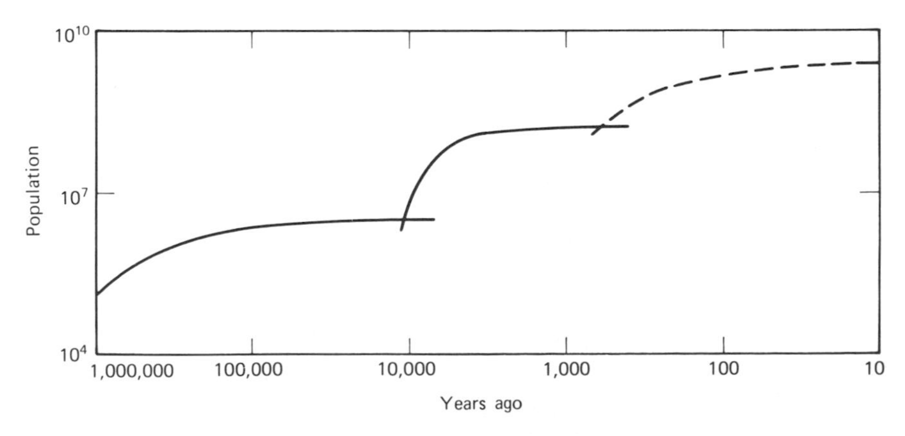

FIGURE 10.2 The history of human population growth. (From Edward S. Deevey, Jr., "The Human Population." Copyright © 1960 by Scientific American, Inc. All rights reserved.

land use, social organization, and political organization; (3) natural catastrophies did not greater alter the population growth rate but caused only minor and temporary reversals; and (4) the rate of human population growth was quite high *prior* to the introduction of health services and sanitation practices by the Industrial Revolution (Spooner, 1972, p. xxv).

The assumption of normally high birth- *and* deathrates is also contradicted by anthropological observations of nonindustrial societies. Many years ago Moni Nag (1962) made a study of 61 nonindustrial societies on which some fertility information was available and found a wide range of variation from very high to very low fertility rates. Particularly important was the discovery that, contrary to the common belief of demographers, fertility rates in nonindustrialized societies were often *lower* than those in industrial societies. Why? Women in hunting and gathering societies are often faced with so much *work* that large families are simply not wanted (Hayden, 1972). Smaller families can be maintained through several cultural practices documented in nonindustrial societies. Infanticide and abortion are common. Increasing the interval between children is another way of maintaining a small family, and this can be done by prolonging the period of lactation for each child, by sexual abstinence, or by a variety of other methods requiring no advanced technology. Childspacing is exceedingly important for nonsedentary hunters and gatherers, since a mother cannot easily handle more than one immature child at a time.

There are two views on mortality rates. One is that while mortality rates are often quite high, they are not sufficiently high to stabilize population growth. There is a general consensus among demographers (and paleodemographers) that 50 percent of all new births died before they reached the age of reproduction. Even so, Polgar argues, it can be assumed that each woman gave birth to eight children during her reproductive period and four of these (50 percent) lived. If two of these are females (and on the average a little less than half of all offspring would be females if the population were large), the net reproductive rate would be 2.0. To put this another way, with a growth rate of 2.0 the population would double in size each generation. Even if you assume that only six children were born to each female, a growth rate of 1.5 is still possible.

The other view is that mortality rates were not consistently high, although they may have been high at times (Kunstadter, 1972). In this case it is assumed that, contrary to popular opinion, mortality due to disease running rampant in the absence of adequate medical services and sanitation facilities is not high. The reasoning behind the assumption is these groups are small and relatively isolated so that epidemic diseases would not be easily spread, camps are kept relatively clean by natural and domesticated scavengers (e.g., the semidomesticated dingo kept by the Australian aborigines), and camps are frequently moved so that people are not living long in their own refuse. Furthermore, hunters and gatherers seldom suffer from malnutrition or starvation, and accidents are not often fatal (Dunn, 1968). However, two points must be kept in mind: (1) The kind and intensity of disease varies with the kind of ecosystem in which the hunters and gatherers participate, and (2) mortality rates due to disease probably jumped dramatically with the sedentism associated with intensified food collecting and with farming. Kunstadter (1972) supports the idea that in many intensive farming societies mortality rates *oscillate* from medium to high because of the periodic occurrence of epidemic diseases. According to this view population grows rapidly during periods when mortality rates are medium but a "crash" occurs whenever epidemics strike and mortality rates zoom upward. The reader will recognize this type of growth as "J-shaped" or "boom-bust," and such a pattern seems to be fairly common among nonindustrialized farming societies.

The incidence of warfare must be mentioned as a possible cause of high mortality. Some feel that warfare is overrated as a cause of death, since it mostly affects males and seldom involves the killing of more than 10 percent of the males of reproductive age (Polgar, 1972, p. 206). However, recent studies of the Yanomamo of Venezuela by Napoleon Chagnon (e.g., 1974) suggests that in some nonindustrial societies warfare can take a large toll. In one village studied by Chagnon, nearly 50 percent of the males were killed in war. A more cautionary statement would be that mortality due to warfare is variable and can be quite high.

MODELS OF POPULATION REGULATION

How are the growth rates of populations regulated to follow one of the growth patterns described above? One possibility is that purely *random* changes in birth- and deathrates are sufficient to explain the balanced growth of some populations, the rapid positive growth of others, and the extinction of yet others. Biologists have investigated this possibility carefully and most have concluded that it is not a sufficient explanation for the dynamics of population growth (e.g., McLaren, 1971, p. 5). A better explanation, they feel, is that some *systematic* process is operating to regulate population size, perhaps in conjunction with random processes. The systematic process provides resistance to population growth. What is responsible for "environmental resistance?" There are two interrelated, though often polarized, sets of processes: density-independent and density-dependent.

Density-Independent Processes

All those factors in the environment that do not operate at a greater intensity on high density than on low density populations are called density-independent. By far the best example are the stochastic (random) occurrences of climatic or geological catastrophes, such as typhoons, earthquakes, and volcanic eruptions, but other factors can be important.

Richard Lee (1972b) has proposed a density-independent model of population growth among hunters and gatherers that centers upon the distances traveled by women and the loads that they must carry. According to Lee's model, if women must carry relatively heavy loads over long distances in the pursuit of subsistence, in visiting other camps, or in moving with their own group, the increased weight added by carrying young children places severe restrictions on the number of children that can be handled at any one time. Therefore, cultural and other mechanisms will be invoked to increase the interval between births, thus keeping the population size low. However, with increasing sedentism, caused by a change in subsistence, warfare, or other factors, the female work load is reduced and more children can be cared for without undue stress. As a result no effort is made to keep birth intervals long and population size begins to increase. Lee uses data on the !Kung Bushmen of the Kalahari Desert to support the model. Traditionally, the !Kung were hunters and gatherers with about two-thirds of their food coming from wild vegetables collected by the women. Vegetable collecting requires considerable movement on the part of the women—3 to 20 kilometers a day for two or three days every week. In addition a woman spends another one or two days each week visiting the other !Kung camps anywhere from 2 to 16 kilometers distant. Finally, camps are moved up to 100 kilometers every three months in the winter and every two or three weeks in the summer. Overall the average !Kung woman travels about 2400 kilometers every year while collecting vegetables, visiting other camps, and moving with the group.

In the pursuit of these activities the !Kung woman carries loads of food, water containers, household utensils, and gifts weighing anywhere from 5 to 15 kilograms. Perhaps more important, the woman must also carry with her children under four years of age. Young !Kung children and their mothers are very close during the first few years and breast feeding is practiced until the third or fourth year. When the mother travels, the child is carried with her, first in a back pouch and later straddling her shoulder. Over the four-year period the child will have been carried about 7400 kilometers.

Lee argues that the distance traveled and the load carried by !Kung women places strict limits upon their rate of reproduction. A child born every four years means that the woman will have to carry only one child at a time. However, if a child is born every two years, the woman must carry two children at once, one very young child in her back pouch and a slightly older child on her shoulder. Lee's data indicate that the great majority of fertile !Kung women have live births every four years on the average, supporting the conclusion that the work load is simply too large if the interval between births is short. The long birth interval is maintained principally by a long period of breast feeding which tends to suppress ovulation, but other cultural devices such as infanticide are also used.

What happens if the group becomes more sedentary? The distance traveled by each woman is reduced, correspondingly reducing the work load. Lee shows that for the Bushmen woman a reduction in annual movement from 1200 to 2400 kilometers to 800 to 1600 kilometers means that a birth interval of three years produces approximately the same work load as a five-year interval did with the greater annual movement. Consequently, a woman can have more children for the same work effort. Direct support for the proposition is found in the Bushmen groups that have become more sedentary in recent years after adopting some domesticated plants (maize, sorghum, and melons) and animals (goats). Their average birth interval has been reduced to three years or slightly less.

Density-Dependent Processes

The second group of regulatory processes includes disease, parasitism, predation, and competition for necessary resources. They are either directly or inversely related to population density. *Behavioral* regulation of population size is the most controversial kind of density-dependent control mechanism. Although not a unique idea, it received an important stimulus from the publication of V. C. Wynne-Edwards' *Animal Dispersion in Relation To Social Behavior* in 1962. At the heart of Wynne-Edwards' theory is the proposition that *group social organization* is essential to maintaining population size *below* the level of the environment's carrying capacity. Social organization includes (1) territoriality, a form of exclusion behavior that operates to divide the resources of a habitat into a limited number of holdings, each of which can supply the group with sufficient food; (2) behavioral limitations on the number of individuals who actually reproduce, for example, restricting nest sites and permitting only individuals with nests to reproduce; and (3) a social hierarchy, in which individuals are ranked in their accessibility to scarce resources, including food and a mating partner, and in which subordinate individuals are not allowed access to scarce resources during times of environmental stress (G. C. Williams, 1971, pp. 9–10).

Territoriality and *group selection* are the most important, and controversial, aspects of the behavioral regulation hypothesis. Group selection has previously been discussed (Chapter 2) so only territoriality will be covered here.

TERRITORIALITY. Human groups have often been observed to partition physical space into portions that are "owned" and sometimes defended from outsiders. Territorial behavior of this sort has been much discussed by biologists and anthropologists alike. Farmers are almost universally territorial, but the existence of territoriality among hunters and gatherers is controversial. Ashley Montagu (1968), for example, has edited a collection of papers with evidence that it does not exist in such groups. On the other hand, Radcliffe-Brown (1948) has observed territoriality in the Andaman Islands, and many authorities have argued for territoriality among Australian hunters and gatherers (e.g., Elkin, 1964; Hart and Pilling, 1960).

Several problems of definition enter into the controversy. In the first place, "territory" is often used synonymously with "area used for subsistence" so that one is never sure whether "home range" or "territory" is the appropriate concept. In the second place, private and group "ownership" of pigeon blinds, fish weirs, and ambush spots on game trails has been documented frequently for hunting-gathering populations, even though a concept of group or private ownership of the habitat containing the owned object is lacking. The Yurok of northwestern California are typical.

Here villages are permanent, house sites are owned individually, and each house is named. Private ownership included such economic resources as sections of ocean beach where surf fishes could be dip-netted, sea mammals hunted, and where occasionally a dead whale might be thrown ashore; groves of oak trees which yielded one of the food staples, acorns; pools where salmon could be speared or netted; and favored spots on deer trails where these animals could be snared or noosed. Yurok diet was based on acorns, salmon, and deer, and Yurok territory held these resources in abundance. Villages did not collectively own or control the economic resources in the vicinity of the settlement–rather, choicest hunting, fishing, and collecting locations were individually and exclusively owned. A highly structured system of property rights and inheritance rules encouraged the development of the Yurok system outlined here. Elsewhere among California Indians, similar individual or family usufructuary rights were known, but the Yurok emphasized this more than any other tribe. (Hole and Heizer, 1973, p. 381)

Finally, while the basic criterion of territoriality has generally been physical conflict or overt defensive behavior, it is well known among ethologists that habitat exclusion is often accomplished by "marking" territories with scents, vocalizations, or other "signs" that are species-specific. Consequently, the occurrence of avoidance behavior among adjacent human populations based upon signs is sufficient cause to infer territoriality. Holmberg, while denying the occurrence of territorial behavior among the Siriono, gives evidence to the contrary:

Bands posses no prescribed territory. If one band runs across hunting trails of another, however, they do not hunt in that area. When I was traveling with Indians of one band in the neighborhood of a house of another, they were reluctant to do any hunting. Informants told me that where trails of another band existed, the animals of that area belonged to the people who made the trails. (1969, p. 132)

Demarcation of territory in this case is accomplished by signs, whether the placing of signs is intentional or not.

Even if a habitat or resource is "owned," boundaries are sometimes crossed with ease by other groups. The concept of *social space* is necessary to understand why a human population can exhibit territorial behavior and still allow some access by other groups. Social relationships between groups are such that some groups interact more frequently and more intensely than others. Such groups are separated by "loose" social boundaries and members may be frequently exchanged, either for purposes of marriage, visits, cooperation in subsistence activities or for a variety of other reasons. As long as the social boundaries between groups are loose, "owned" physical space becomes more a theoretical concept than an expression of actual behavior. Consequently, resources or land belonging to one group can be used by others if the social relationships are sufficiently close to make the outsiders practicing, if not actual, members of the group. Of course, social boundaries may change and groups that are the same physical distance apart may feel that they belong to the same group at one time and very different groups at other times.

The ecological expediency of "loose" social boundaries for mobile hunters and gatherers is suggested by several studies. Richard Lee's data on the !Kung Bushmen show that fluctuations in the availability and distribution of water are responsible for cycles of aggregation and dispersion of local groups (1972a). The !Kung are organized into small local camps, a noncorporate, bilaterally organized group living in a single settlement and having at its core 2, 3, or more adult males and females who are siblings or cousins. Each camp occupies, and considers itself to "own," a habitat centered upon one or more water sources and surrounding resources. During periods of maximum water availability the camps are widely dispersed, taking advantage of permanent water holes, semipermanent water holes (those that fail occasionally), and a series of temporary water holes that are available from a few days to a few months. Figure 10.3a shows the reconstructed distribution of 11 !Kung camps during the wet seasons and good rain years from 1920 to 1930. During dry seasons and periods of drought many water holes dried up and forced their !Kung "owners" to leave. Figure 10.3b shows that the dispossessed camps moved to more permanent water holes and temporarily joined the !Kung that "owned" these water holes. It is in this sense that social boundaries are considered to be loose. Members of one camp are readily accepted into other camps when the need arises.

David Damas (1969) provides data from the Central Eskimo showing a similar ease of movement across social boundaries. During the winter, groups of 50 to 200 people aggregated for purposes of *maultiqtuq* or breathing-hole seal hunting. A large group of seal hunters is required for this kind of activity, since an individual seal maintains several breathing holes in the ice over a wide area. One or even a few hunters could not keep constant watch over all of the holes to harpoon the seal when it surfaced to breathe. During the rest of the year the winter aggregation dispersed into small groups of 15 to 20 persons to fish in lakes and to hunt caribou, small game, and birds inland. Occasionally, during the summer, groups of 50 or more persons aggregated for purposes of weir fishing during char migrations and the hunting of migrating caribou.

In summary, then, there is little anthropological evidence suggesting a genetic propensity for human populations to be territorial. Hunting and gathering groups frequently have concepts of land or resource "ownership," but social boundaries between groups are so loose that owned territory is easily used by neighbors. Indeed, loose social boundaries are essential to survival in habitats with fluctuating resources. Farming groups that depend upon plots of land to produce resources year after year, on the other hand, are much more inclined to have well-developed concepts of territory with tighter social boundaries between groups. There is some evidence of *individual spacing* behavior in all human groups. For example, the psychologist Robert Sommer (1969) shows a number of experiments that personal space is very much a part of modern American life. Americans are inclined to take the same seat in a classroom day after

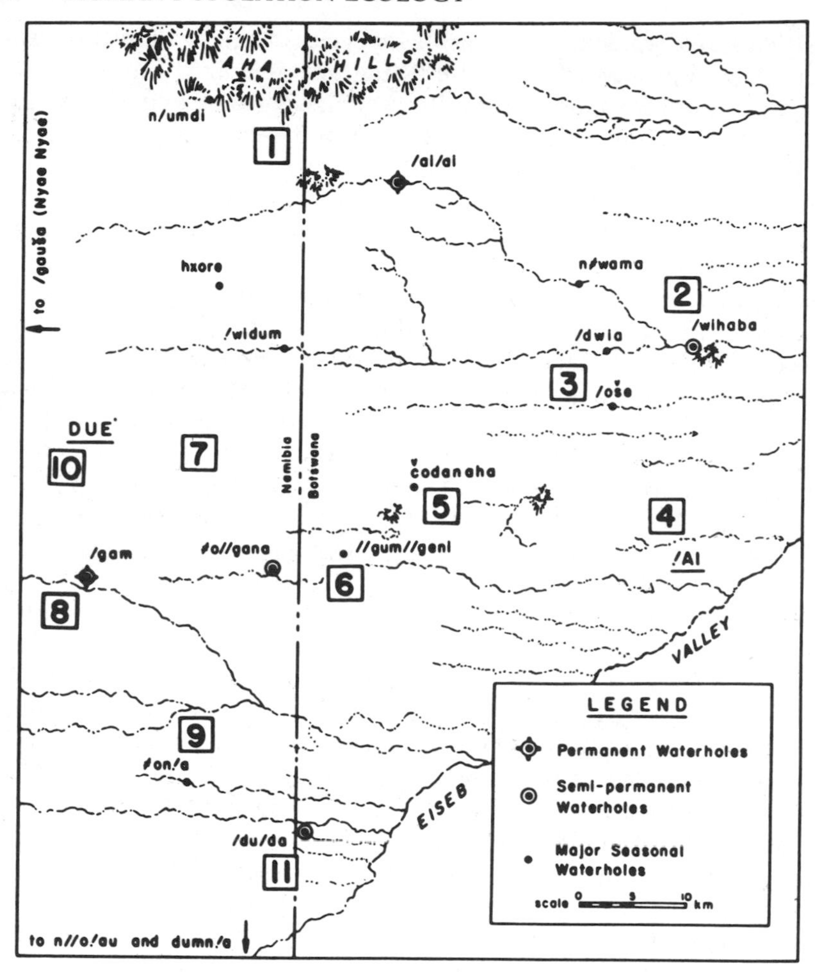

FIGURE 10.3a Seasonal changes in the distribution of !Kung Bushmen camps: Distribution during the wet season and good rain years. (From Richard Lee, "!Kung Spatial Organization: An Ecological and Historical Perspective," *Human Ecology*, 1, 1972. Copyright © 1972 by the Plenum Publishing Corporation. Reprinted by permission of the publisher.)

day, or to sit at the same table, and feel annoyed if their position is usurped. They maintain an invisible volume of space around their person and feel uncomfortable, if not downright aggressive, if that space is unexpectedly penetrated by others. Studies by the anthropologist Edward Hall (1966) have also shown that the amount of personal space needed varies from one human group to another and that it appears to be culturally learned. Finlanders, for example, require more space than do Americans, and Americans require more space than do Latin Americans. Nevertheless, we do not know whether there is an upper and a lower limit on personal space that is established through genetic inheritance.

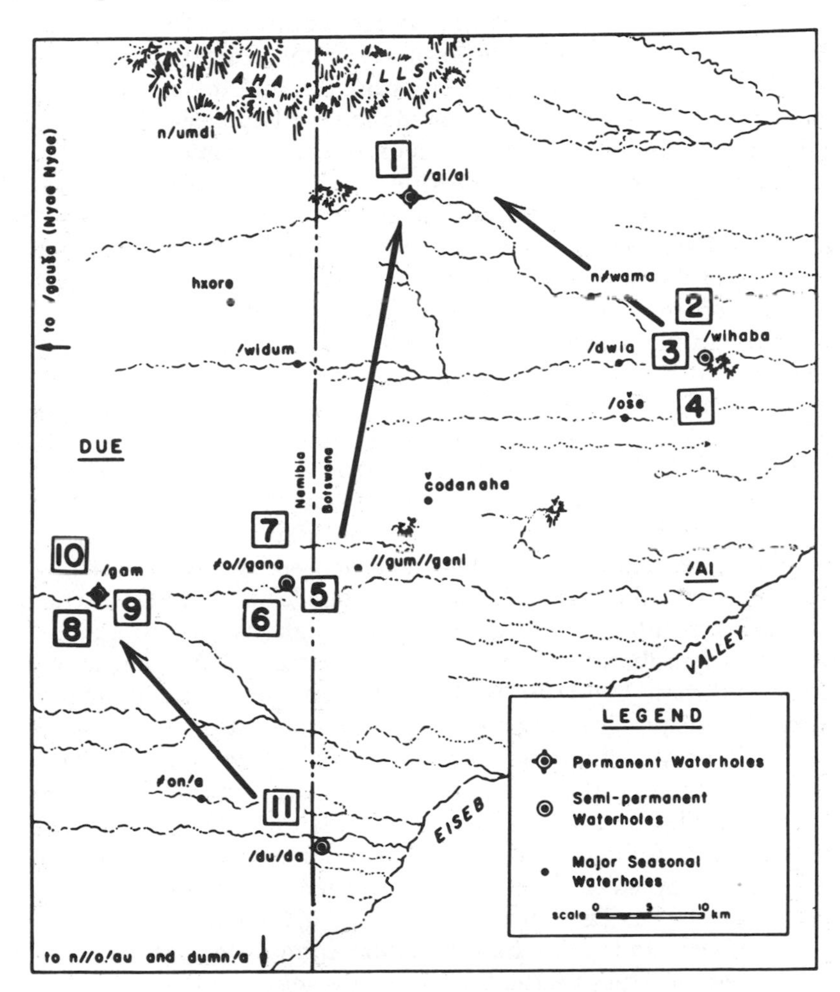

FIGURE 10.3b Seasonal changes in the distribution of !Kung Bushmen Camps: Distribution during the dry season and drought years. (From Richard Lee, "!Kung Spatial Organization: An Ecological and Historical Perspective," *Human Ecology*, 1, 1972. Copyright © 1972 by the Plenum Publishing Corporation. Reprinted by permission of the publisher.)

OTHER DENSITY-DEPENDENT PROCESSES. Brian Hayden (1972) proposes a variety of density-dependent models to human populations based upon sexual dominance behavior regulated by the scarcity of foods and, as in Lee's model, the amount of *work* required of women. Dominance behavior has been observed in many primate societies and is particularly marked under conditions of environmental stress, for example, when food is in short supply. The result is increased aggression by dominants toward subordinates, and it is to be expected that mortality rates within the population will increase accordingly. Hayden points out that in primate groups females are generally subordinate to males. Increasing aggression due to environmental stresses would cause increased mortality not only in female members but also in the infants closely associated with the females. The subsequent scarcity of females in the population combined with high infant mortality rates could be an effective method indeed of controlling population size.

In human populations, of course, there is no direct evidence that male dominance causes higher mortality rates among females during times of environmental stress. However, a dominance hierarchy based upon sex is essential to set into motion institutionalized ways of regulating population size *that are controlled by women.* These include abortion, abstinence, prolonged lactation, and infanticide. Hayden argues that the dominance hierarchy is used to divide the population into a *working group* consisting largely of mature women and a *nonworking group* consisting of immature young, the very old, some religious leaders, and particularly *males.* Males in hunting and gathering populations, according to this view, really contribute very little to the fundamental food needs of the group, and Richard Lee's (1972b) observation that the males in the !Kung Bushmen bands provide only one third of the total subsistence supports the idea. (However, this model must be reversed for some specialized hunters such as the Eskimo; these males provide almost all the food.) Restricting the labor supply in this way limits the size of the group by placing pressure upon the women who are doing most of the work. As the size of the population increases, the women simply do not have *time* to raise more offspring and they cannot afford to feed more mouths. As a result they begin to invoke the population control devices available to them, as mentioned above. Regardless of how much work is done by the working group, if it remains small in relation to the entire population, the average per capita work load will be low; this seems to be characteristic of most hunters and gatherers. However, among intensive farmers the size of the working group increases tremendously and the per capita work load is quite large (M. Harris, 1975).

The participation of women in the labor force has already been mentioned as a possible cultural factor reducing fertility in twentieth-century America. However, in this case *both* men and women are working, and the per capita amount of work is high. Fertility rates, then, can be reduced simply by bringing women into the labor force in such a way that the time necessary to bear and raise children is greatly reduced. Whether men are working or not working seems to have little effect upon fertility. The importance of Hayden's model is that it shows how the size of a population can be kept well below the carrying capacity of the habitat. If both men and women of all ages were working, the group would have a much greater impact upon the natural resources used for food, an impact that in bad years would strain the ability of the ecological system to maintain these resources at a level adequate for the size of the group.

A density-dependent model that is related to behavioral regulation and has gained popularity in recent years is the regulation of population size and density by *stress*. The model is based upon experimental evidence that overcrowding stimulates social stress that, in turn, has deleterious effects upon the individual because of feedback from the endocrine system. In higher animals most of the endocrine feedback originates in the pituitary and adrenal glands and is believed to intensify processes that decrease production, increase mortality, or both. Observations of rats and rabbits under laboratory conditions and various mammals under natural conditions suggest that if population density is allowed to increase, a point is reached when drastic behavioral and physiological changes take place. Most notable are increased aggressive or other antisocial behavior and/or a cessation of breeding. The growth of individuals is inhibited and the size of the adrenal gland increases in relation to body size.

These observations have been projected to humans to explain the occurrences of antisocial behavior and other social pathologies in high-density urban areas. However, there is little direct evidence that such a relationship actually exists. The psychologist J. L. Freeman (1973) has summarized the evidence from demographic studies, field and observation studies, and experimental studies and concludes that it is weak and inconclusive. Demographic studies show that there are consistently higher rates of crime and mental illness in cities than in the country and in city centers than in city borders. However, it is impossible to correlate these rates with densities because of the confounding effects of other social and economic factors. Thus Los Angeles has a higher rate of major crime than New York but a lower population density. There is a positive correlation between juvenile delinquency and population density in Philadelphia and Denver but a negative correlation in Chicago. In areas where strong correlations have been found between density and social pathology, factoring out such socioeconomic factors as education and income have greatly reduced the strength of the correlations. The overall picture is confusing. Cities have higher crime rates than rural areas, but low-density cities often have higher crime rates than high-density cities. Central cities and slums have high crime rates but this is more often correlated with education and income rather than with density.

Field studies are based upon detailed studies of individuals in actual situations. An early study in Paris suggested that the amount of apartment space available per individual was correlated with the incidence of social pathologies. The incidence increased greatly beyond both an upper and lower limit on living space. However, a later study in Hong Kong with many more subjects suggested no correlation between space and aggressiveness and emotional stress. Other studies have been done on the crowding of children in playground space and have found little evidence of density effects. However, these studies do show that larger groups of children at any density are more aggressive and less social.

Finally, there are a few relevant controlled experiments. Most of these are studies of the effect of isolation upon the individual, appropriate here because the lack of privacy in high-density situations is supposed to be the most immediate cause of emotional stress. The majority of the experiments are concerned with the ability of subjects to perform various tasks under conditions of isolation or crowding. The results show that if the isolated groups were *all male,* the subjects tended to show emotional stress and reduced ability to perform tasks under crowded conditions. However, if the groups were mixed *male and female,* the subjects showed no stress or reduced ability at all. Furthermore, if the groups were *all female,* the subjects performed better under crowded conditions.

SUMMARY

Classic demographic theory sees the growth of human populations as following an S-shaped pattern. Growth is slow initially because high mortality rates cancel out high fertility rates. A population explosion takes place when high mortality rates are reduced, often because of the introduction of medical services. Finally, population size levels off when fertility rates drop to achieve a new balance with mortality. Several anthropologists have challenged this picture, pointing out that fertility and mortality rates are quite variable in all human populations, including preindustrial ones, and that moderate rates seem to be more appropriate for observed hunters and gatherers. The variability in demographic rates gives all human populations the potential to grow rapidly and decrease just as rapidly, suggesting that J-shaped growth is a common pattern.

Recognition of ecological relationships affecting fertility and mortality rates is critical to understanding the actual pattern of population growth. Models of population regulation are based upon such relationships. Density independent models assume that population size is regulated by physical, social, cultural, and biological processes that operate independently of how large or dense the population is. By contrast density-dependent models assume that the intensity of such processes in regulating size is affected by size itself. In reality the size of human populations is probably controlled by both kinds of mechanisms. George Cowgill gives a useful summary of this perspective:

Rather than seeing population growth as an inherent tendency of human populations which is permitted *[italics in original] by technological innovations, I see population growth as a human possibility which is* encouraged *[italics in original] by certain institutional, as well as technological or environmental circumstances, but equally may be* discouraged *[italics in original] by other circumstances. In general, asking whether population growth is "the" independent or "the" dependent variable is an inept question, and we should think of population variables as members of sets of variables, including technological and environmental variables and political, economic, and other institutions, which are all concomitantly interacting with one another. (1975, pp. 516–517)*

ELEVEN CARRYING CAPACITY

Many of the models of population growth used in anthropology assume that there is a limit beyond which further growth is inhibited. The best example is the S-shaped or logistic model. Growth takes place at an exponential rate until increasing "resistance" from density-independent and density-dependent factors forces it to level off at some upper limit, usually symbolized by a constant K and called the *carrying capacity*. In theory the carrying capacity is the size at which a population can be permanently supported by its environment. Models with growth limits of this kind are usually called *Malthusian* in recognition of Thomas Malthus, the English clergyman who first proposed that resources available to a population set rigid limits on its growth.

However, it has been pointed out that the designation *Darwinian* is more correct, since Malthus later changed his mind and Charles Darwin immortalized the idea in his concept of natural selection (Peterson, 1975). Darwinian models of growth are necessarily static and focus on the ways that a human population keeps its size within the limits of material support. In recent years the interest of demographers in the problems of economic development, particularly in Third World, had led some to reject the static Darwinian model in favor of *population pressure* models that see population growth as a "prime mover" in bringing about social and technological change. The concept of "limits" or carrying capacity is still there, but as the size of a population approaches the limit social and technological innovations are stimulated to relieve the pressure.

Anthropologists have adopted both Darwinian and population pressure models for use in ecological studies. Darwinian models are mostly used to estimate the size of now extinct populations. Population pressure models are used to explain why ecological change takes place. The concept of carrying capacity is essential to both kinds of models, so let us examine its use, and problems arising from that use, in detail.

LIMITING FACTORS

The concept of carrying capacity is derived from the idea that an organism can exist only within a limited range of physical conditions. It must have access to at least a minimum amount of energy and critical materials; it must have certain temperatures; it cannot survive certain concentrations of chemicals; and so forth. The availability of suitable conditions for living determines how many of an organism can exist in an environment.

In 1840 the plant physiologist Justus Liebig showed that organisms will be limited by materials that are in the shortest supply relative to their need, a relationship known today as the *Law of the Minimum.* The ecologist Victor E. Shelford (1913) carried Liebig's idea one step further during the early part of this century by proposing the "law of tolerance." Shelford argued that animals could survive only under conditions that they could tolerate and that the availability of critical materials is only one of those conditions. Others include temperature, sunlight, and hazards.

The classic study of limiting factors on human populations was made by the anthropologist Joseph Birdsell (1953). Birdsell shows that water availability is closely correlated with the populaton density of aboriginal hunters and gatherers in Australia. The study is based upon the assumption that in a desert area with little regional variation in temperature and elevation, such as Australia, the abundance of plants is determined by how much water there is. Plant biomass, because of food pyramid relationships, is further assumed to reflect the food supply available to hunters and gatherers and, therefore, to be closely correlated with their abundance. *Mean annual rainfall* is used as an indicator of available water, but Birdsell takes care to eliminate from his sample of Australian hunters and gatherers those that received "unearned" water, for example, water coming from rivers with a drainage basin *outside* their territory. He assumes that the ecological population of the Australian aborigines is identical to the "tribe" and that its size is always 500 individuals. While there is some variation in size, Birdsell's statistical examination of data on a large number of tribes supports his assumption that the mode or central tendency is indeed around 500. Working from this assumption, Birdsell proposes that the size of the area occupied by each tribe, and therefore tribal population density, is a function of mean annual rainfall. To test the hypothesis, a planimeter is used to measure Australian tribal areas on a map published by Norman B. Tindale. These areas are then plotted against the mean annual rainfall averaged over the entire tribal area. Rainfall values are calculated from maps given in a climatological atlas of Australia. Figure 11.1 shows the results. The curve is exponential of the form $Y = aX^b$, in which Y is the tribal area and X is the mean annual rainfall, and is based upon Birdsell's "basic series" of 123 tribes, selected to eliminate those that receive "unearned" water, make use of ocean foods, and have unusual cultural characteristics. Mean annual rainfall and tribal area show a correlation of $r = 0.81$.

There is some question, however, that average rainfall is the best measure of limitations upon plant growth. Soil properites, for example, determine how much rainfall can actually be used by plants. Consequently, plant growth in two areas with an identical mean annual rainfall can be radically different if soils are dissimilar. For this reason a measure that is based upon "effective" water, that is, water actually available to plants, is likely to be a better indicator of plant growth. *Actual evapotranspiration* (AE) has been proposed as such a measure (Rosenzweig, 1968). Actual evapotranspiration is the amount of water that evaporates from the land surface in the form of runoff or percolation plus the amount of water that transpires from the stomata of plants. Preliminary studies have shown that AE is closely correlated with *net primary productivity.* A recent reexamination of Birdsell's basic Australian series has taken advantage of this relationship and has shown that AE is a better predicator of tribal area, and therefore population density, than mean annual rainfall (Casteel, 1974).

The concept of limiting factors must be used with care. Paul Colinvaux (1973, p. 277) points out that "thinking of the environment as set of limits has practical use, but the impression given by the meaningless 'laws' of the minimum and tolerance, that single factors were often limiting, is false." Yet, as in Birdsell's study, most applications of the concept in anthropology have looked only at single limiting factors. In reality the limits are much more complex, as suggested by recent work by ecologists on Liebig's principle (E. Odum, 1971, pp. 106–107). First of all, other factors in the environment that are in abundance may actually moderate the effects of a potentially limiting factor. Second, the Law of the Minimum may not hold when the ecological system is unstable and undergoing change. When this happens, it is not the least abundant element that is limiting but the interaction among many elements. The problem with the concepts of limits and tolerance is that

they have conveyed the idea that all you had to do in your field studies was to go out and measure things until you found something which seemed to be limiting, and that you had then explained something significant. The chances were that you

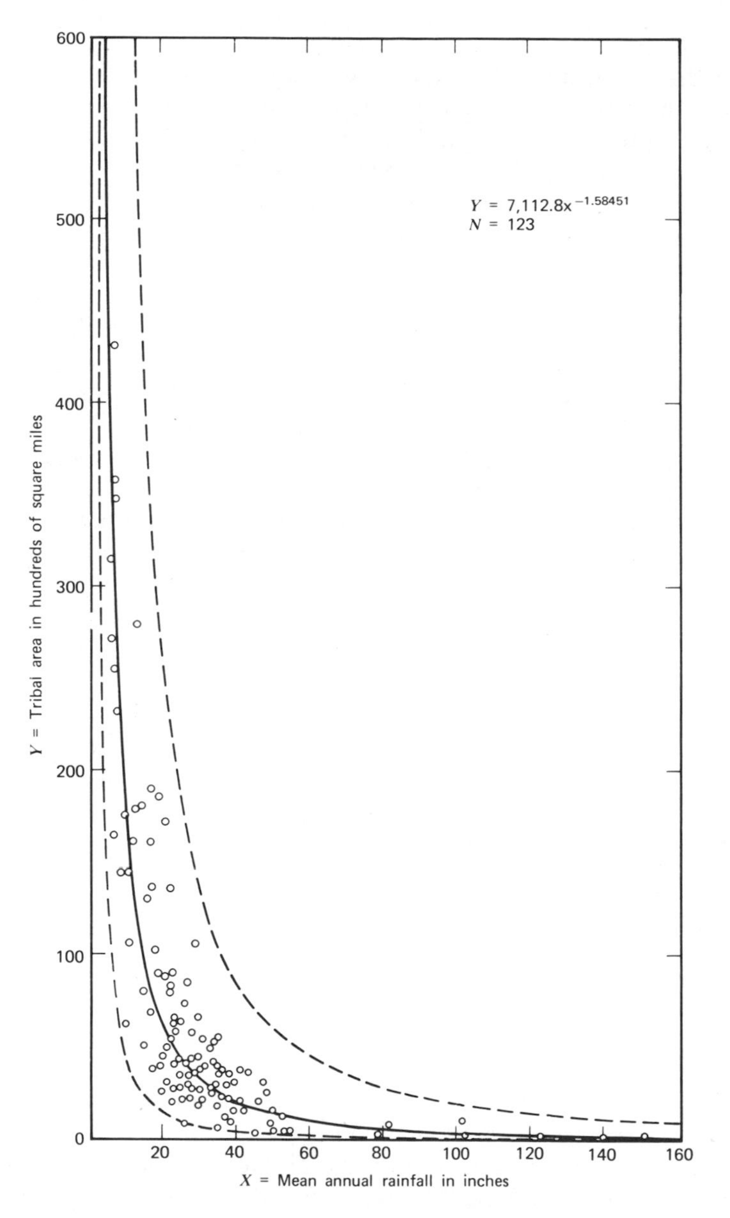

FIGURE 11.1 The correlation between mean annual rainfall and the size of Australian Aborigine tribal areas. (From Joseph B. Birdsell, "Some Environmental and Cultural Factors Influencing the Structuring of Australian Aboriginal Populations," *American Naturalist*, 87, 1953. Copyright © 1953 by the University of Chicago Press. Reprinted by permission of the publisher.)

had done nothing of the sort. You may have come across some temporary conditions in which an animal had bumped against some real limit or even that the correlation with an apparent restraint was entirely fortuitous. (Colinvaux, 1973, pp. 276–77)

I cannot reiterate too many times that *complexity* is the *sine qua non* of ecological studies and that simple correlations will seldom give the whole story.

METHODS OF ESTIMATING CARRYING CAPACITY

The principal problem with attempts to estimate the carrying capacity for human populations is precisely the use of a single limiting factor–the amount of land required to grow, hunt, or gather enough calories to feed one individual. Such a method not only ignores two fundamental problems of life—acquiring critical materials and coping with hazards-—but also ignores human symbolism. A plot of land may have a low carrying capacity, not because of low soil fertility but because it is sacred or inhabited by ghosts. Furthermore, even if it is assumed that carrying capacity can be accurately estimated from calories, conventional methods are subject to numerous other problems of measurement and assumption. Perhaps the best way to point out the problems is to go through one method, that used to calculate the carrying capacity for farming groups, step by step and to closely examine each variable that is used. The method requires (1) estimating the total amount of land that is available for cultivation, and (2) dividing by the amount of farming land that is needed to support each person. Both estimations make a number of assumptions. The assumptions in estimating per person land need are particularly numerous and difficult to justify, thus introducing many sources of error into the calculation of carrying capacity.

Total available land is a function of technology and habitat and varies greatly from one area to another. In tropical Afica, for example, the geographer W. Allan (1967) gives estimates ranging from 5 percent of the total land area to around 50 percent, with an average figure of 25 percent. The reasons why land cannot be used are numerous and include both social and physical environment: endemic diseases, warfare, insufficient or excessive water, steep slopes, poor fertility, sacredness. It is also clear that technology is a prime force in determining what land can be used and what cannot be used. For example, the steep mountain slopes of the Peruvian Andes could not be used for farming without significant modification but were used by the ancient Peruvians after the introduction of terracing technology. Similarly, arid lands are made suitable for cultivation with irrigation technology, in the same way that swampland can be reclaimed or an advanced medical system can be introduced which renders livable lands that were formerly uninhabitable because of endemic disease. All in all, a reasonable estimate of land available for farming depends upon an accurate assessment of technology, physical habitat, and social environment.

Individual land need is based upon several factors, including the food requirement of each person, the productivity of the farming land, and the intensity of land use. Most difficult to estimate is individual food need, which can either be based upon *actual* consumption or upon experimentally derived values. Both approaches are hazardous. Actual consumption may not be an accurate measure of what is required in the long run. (We are all familiar with "unhealthy diets.") Presumably, however, actual consumption is a good approximation if the population is judged to be medically healthy. Perhaps as important is the difficulty in observing what is being eaten (people do not always eat in the open and they are often reluctant to let strangers watch) and in obtaining permission to collect essential data about the food (e.g., weight). If nothing else, the time required is enormous. Typically, detailed observations and measurements are made on a small segment of the group, perhaps a lineage, and the results projected to the entire population. Experimentally derived estimates of food needs are usually based upon Western food experiences and seldom reflect genetic variation in food needs.

For horticulturalists Roy Rappaport (1968, pp. 288–92) gives several methods of calculating the amount of farming land required to adequately feed each person, based upon data for the Tsembaga Maring of the New Guinea Highlands. One method is to estimate the calories consumed by the entire population during a single year and to divide that figure by the annual caloric yield of each acre of farmland. Group consumption is calculated by counting the number of persons in each age and sex category, estimating the calories needed by a person in each of the categories, multiplying the two, and then adding the products. The average caloric requirements for an individual of a particular age and sex category are based upon detailed observations of the actual consumption by members of a single clan, supplemented by data from an earlier nutritional survey of the New Guinea Highlands. The second step is to calculate the amount of land that must be put into production to supply the calories needed by the Tsembaga for each year. Estimates are made of the yearly calorie production of an average acre of farming land. This figure is divided into annual group consumption (minus any calories that may have come from sources other than the farming land used in the calculation) to obtain the number of acres necessary to adequately feed the entire population. The third, and final, step is to divide the population size into the number of acres required to support the population to get the land needed to feed each person.

What about differences in the intensity of land use? Intensive farmers use the same land year after year so that the annual land need will be identical to the long-term need. However, shifting farmers do not use the same land but move to new fields after one or more years. Consequently, the land needed for any one year is only a part of the land needed for permanent support. To correct for differences of this sort, a "land-use" factor is included in the calculation of carrying capacity. The land-use factor is of the form

$$\frac{R + U}{U}$$

where R is the number of consecutive years that farming land is allowed to rest or fallow, and U the number of consecutive years the farming land is used. If the land is used continuously, as in intensive farming, the value of R is zero and the land-use factor becomes unity or 1. However, if the land is rested at all, the value of R is greater than zero and the land-use factor becomes greater than 1. For example, if the field is used for 2 years and then rested for 18 years, the land-use factor is

$$\frac{18 + 2}{2} = \frac{20}{2} = 10$$

In this case the individual needs 10 times as much land for permanent subsistence as for yearly subsistence.

Now let us proceed to the actual calculation for carrying capacity. The basic equation is as follows (as modified from Carneiro, 1960):

$$\text{carrying capacity} = \frac{\text{total available land}}{\text{individual annual land need} \times (\text{land-use factor})}$$

For example, suppose that the total available land is 1000 acres, the individual annual land need 2 acres, and land-use factor 10. The carrying capacity is then

$$\frac{1000 \text{ acres}}{(2 \text{ acres/person}) \times (10)} = \frac{1000 \text{ acres}}{20 \text{ acres/person}} = 50 \text{ persons}$$

Joseph S. Weiner, a human biologist, argues that a term for "working capacity" should be included in the calculation of carrying capacity (1972, pp.405–406). Working capacity is the extent to which the individual is biologically able to work. For tropical areas Weiner lists four factors that affect an individual's working capacity:

1. Adequacy of diet. A person is unable to work at peak capacity if caloric, protein, or other food intake is below a minimum level.

2. Intensity of work. Farming labor is much harder under some circumstances than under others; for example, if the terrain is rugged, the effect of poor diet or disease upon working capacity is greater than if the terrain is smooth and level.

3. Intensity of heat load. High temperatures can greatly reduce working capacity; the amount of reduction depends upon the degree to which genetic adaptation and physiological acclimatization to heat have taken place.

4. Disease. Endemic diseases can severely reduce working capacity, particularly those diseases that strike during peak labor periods.

Any reduction in a person's working capacity affects the productivity of the land that is farmed. Consequently, actual productivity figures will be lower than those expected if the individual is working at top capacity. Just what this correction is depends upon the population and the environment, of course. Nevertheless, some adjustment should be made if there is evidence that the working capacity of the members of a population is greatly reduced.

In addition to the problem of working capacity this method of estimating carrying capacity has received other criticisms. The geographer John Street, for example, has outlined weaknesses in four assumptions (1969). Two are interrelated: Farming technology does not change, and there is no change in the crops used for subsistence. Numerous studies exist to show that farming populations, even traditional ones, accept new tools, methods, and crops and that these innovations can dramatically change both available land and land productivity. If constancy in time is subject to criticism, spatial uniformity is even more susceptible. Few studies have supported the idea that farming techniques are identical on all the available land, yet this is a common assumption made in the calculation of carrying capacity. Consider the farming practices in the Valley of Oaxaca in the Central Highlands of Mexico. According to Kent Flannery and his colleagues (1967) the earliest farmers (1500 to 600 B.C.)

made use only of the valley floor near the rivers where a high water table permitted shifting dry farming. In addition they used "pot-irrigation" in the high water table zone by digging shallow wells and irrigating small gardens with pots of water drawn from the well. With the rise of towns and ceremonial centers (600 to 200 B.C.), more advanced farmers continued to use the traditional methods but added two others. Canal irrigation was established by digging canals from high river tributaries down to valley land with a lower water table. Also, for the first time, the foothills of the valley were farmed using the technique of *tlacolol* or *barbecho* (simple dry farming with a fallow period). This combination of intensive farming with irrigation and extensive dry farming in the hinterland is called the "infield-outfield" system after a similar method used in Europe. Later developments in the Valley of Oaxaca were based upon the four basic techniques. Each of the farming methods is associated with different productivities, and there is little doubt that simply averaging the yields would give a wrong estimate of carrying capacity.

Now let us return to Street's remaining criticisms. The land-use factor is an important part of the calculation of carrying capacity, yet there is little if any effort made to be sure that the factor is correct. According to Street the factor is almost always based upon *present* land use, even though there is frequently evidence that the land is deteriorating because of that use. The investigator must be especially careful that the rest and use intervals used in the calculation are sufficient to maintain the land without permanent deterioration. Street's final criticism is the assumption that losses in crop productivity due to weeds, pests, and crop diseases remain constant and do not have to be considered as a factor in the calculation of carrying capacity. Yet studies have shown that as the intensity of land-use increases, that is, as the land-use factor approaches unity, losses not only increase progressively but also undergo "quantum leaps." Correspondingly, the amount of land actually required to support a single person tends to be larger than would be expected.

INSTITUTIONAL ARRANGEMENTS AND CARRYING CAPACITY

Carrying capacity is often assumed to be affected only by human technology. Yet there is abundant evidence that the *institutional arrangements* of a society, its political, kinship, and ritual organization, for example, have an equally strong impact. Consider the following example. Data given by Robert Adams (1965) for the Lower Diyala Basin in Iraq suggests an oscillating pattern of population growth with increasing amplitude through time. The first period of settlement in the basin (400–500 B.C.) is associated with small-scale irrigation with little impact upon the physical environment. Small strips along streams are flooded by cutting through natural levees or by digging small branch canals into the strip. Population growth during this period is marked by fluctuations, but the increases and declines take place at a slow rate and are rather uniformly spread over all the valleys in the basin. The magnitude of the oscillations is on the order of 30 to 40 percent. Irrigation networks are rapidly expanded during the second major period of settlement; by the Sassanian Period (A.D. 226–637) nearly all of the arable land in the Diyala Basin is brought under cultivation. Large feeder and branch canals are dug and the landscape is greatly modified. There is no indication that the land is used more intensively (as would be predicted by the Boserup hypothesis), but there is evidence that measures are taken to bring more land into production. During this period a strong central government is engaged in constructing, operating, and maintaining the large irrigation systems extending into virtually every part of the Diyala Basin. It is also during this period that population surged to several hundred thousand. However, the population "explosion" is followed by a series of large oscillations closely tied to the fortunes of the state government.

The political instability associated with changes in dynasties is often responsible for drastic reductions in population size, apparently because the irrigation networks could not be adequately maintained without a strong, well-organized central government. Silting of the canals is a particularly bothersome problem, and entire valleys often have to be abandoned for that reason. However, with the advent of a new government, canals could once again be maintained and depopulated areas reoccupied. The end of the Sassanian Period is followed by a population "crash" during which nearly half of the total settled area in the Diyala Basin is abandoned. Collapse of the Sassanian state is largely responsible, but the direct reasons for abandonment of a valley are varied: failure to maintain dykes in one region, silting of canals in another, the departure of forcibly settled foreigners with evacuating armies in yet another. However, an overriding factor in nearly all abandoned valleys is the failure of the central state government, a government that is necessary to maintain the vast irrigation network upon which rested the subsistence base. In short, the "carrying capacity" of the region varies with the efficacy of the political organization.

The interaction between *ecological variation* and political organization also varies. In areas with poor water drainage the expansion of the irrigation network results in a drastic rise in the water table to the point where salt accumulates in the top soil through leaching. Increased salinity reduced the "carrying capacity" of the land to nearly zero; even today these areas are unoccupied. Populations in these valleys crashed because of the efforts of a strong central government to increase the carrying capacity of other valleys through water transportation. However, salinification did not take place in valleys along the upper course of the main river system. A much steeper gradient existed and irrigation water drained sufficiently. Consequently, populations did not crash in these regions, but several valleys were abandoned rapidly for other reasons.

In addition to political organization, *land tenure rules* can create differences in carrying capacity in the face of identical habitats and technology, particularly because such rules affect the *geographical distribution* of individual landholdings. William Douglass (1971) gives data showing why this is true for two Spanish Basque villages: Echalar and Murelaga. Echalar farmers have landholdings covering a single territorial entity. Each crop is planted in one large field rather than in several smaller ones; as a result the fields can be worked more efficiently. For example, long furrows can be used when the field is ploughed, greatly reducing the number of times that the team of draft animals must be turned around. By contrast, because of land tenure differences, the otherwise identical village of Murelaga has individual landholdings that are widely separated in small parcels. A single field crop, for example wheat, must be planted in several fields, often far apart, and worked by laboriously moving draft animals and their implements from field to field; even in the same field they must be frequently turned around, a time-consuming and energetically expensive operation. Since the per unit area production of the fields in the two villages is equivalent, the greater investment of labor required in Murelaga reduces its carrying capacity.

Finally, estimates of carrying capacity have ignored the integration of local economies into regional, national, and international market systems and have assumed a *subsistence* economy. Yet, in today's world isolated subsistance economies are few and far between. *Cash crops* are grown everywhere, taking up land that traditionally was used only for food and making the grower susceptible to unforeseen fluctuations in market prices. Market forces are such that a bumper crop year after year does not assure its easy conversion into currency nor does its conversion into currency assure that foodstuffs can be purchased. Tobacco, sisal, and opium cannot be eaten if the market fails; if the market is good, a fortune in cash cannot be used to buy the food that no one has raised. Once again the intricate web of ecological relationships cannot be forgotten. The proportion of land used for cash cropping, the proportion of land used for subsistence crops, weather patterns, market forces—all of these and others interact to determine how many people with a given technology can be supported by the habitat. One thing is certain: Carrying capacity is *dynamic,* changing with changes in a multitude of other factors. It is not the static concept suggested by the logistic population growth curve and by most people who have used it.

TIME AND CARRYING CAPACITY

The dynamic view of carrying capacity is expressed in a recent article by Brian Hayden (1975). He argues that the traditional concept of carrying capacity is weak because the resources in a habitat are not constant; their quantity, and possibly quality, fluctuates over time, often in *cycles.* Consequently, estimations of carrying capacity would have to consider recurrent changes over time in a resource base not only during a year, but also during decades, centuries, and millenia. Hayden believes that a population uses its available biological and cultural devices to adapt to these cycles, or other time trends, rather than to averages. Correspondingly, the goals of demographic studies of this kind should be:

1. Definition of the kind and length of time cycle or trend to which a population is adapted.

2. Understanding of the circumstances surrounding a *change* in adaptation from one kind of cycle to another.

3. Devising methods of measuring adaptation in these terms.

"More work needs to be done" is the advice offered for the first two goals, but Hayden gives a concrete suggestion for the third. He proposes that a useful measure of adaptation is "the frequency intervals with which given intensities of protein-calorie morbidity, or alternately mortality, occur in a given population due to resource shortages" (1975, p. 16). The intensity of protein-calorie malnutrition can be estimated from direct observation, historical records, or in an archaeological context, by using osteological indicators such as "Harris lines." Presumably, if a population undergoes intense stress of this sort quite frequently, it is not well adapted to its resource base. Conversely, a well-adapted population should have long periods of time between intense stress.

POPULATION PRESSURE

Population pressure models have been used by anthropologists to explain everything from the origin of farming to economic development.

The archaeologist Lewis Binford (1968), for example, argues that growth in Mesolithic hunting and gathering populations occupying the Levant coast of the Middle East forced migration of "splinter" groups into the interior. The interior was dominated by microenvironments with a relatively low carrying capacity, ultimately creating a "tension" zone in which population pressure was sufficiently intense to cause a shift away from hunting and gathering to increased dependence upon a new source of food—domesticated plants.

Flannery (1969) gives data from western Iran supporting the model. In this area the Upper Paleolithic period was marked by the hunting of ungulates. However, at the very end of the Paleolithic period ungulate hunting was being supplemented by small game, fish, turtles, water fowl, partridges, snails, mussels, and crabs, among other things, a shift that continued to intensify until well into the Mesolithic about 7500–800 BP. Flannery suggests that the cause of the shift was population growth caused by migration from the Levant region leading to overhunting of ungulates. As a result people began to turn to more dependable resources at some times of the year, particularly migratory water fowl, riverine products, and plant foods that can be stored such as acorns and pistachios. Broadening of the subsistence base was associated with a division of labor, placing more emphasis upon the role of women in gathering resources around a home camp while men continued to hunt ungulates from temporary camps. Domesticated plants appear during the same period as part of the economic shift, and Flannery argues that domestication may be closely associated with the movements of people into more and more marginal areas as population pressure increased. Farming could be used as a way of duplicating in poor areas the stands of wild wheat and other cereals that continue to grow naturally in more fertile areas of western Iran. Experiments by Harlan and Zohary (1966) have shown that stands of wild wheat are so dense in these "optimum" zones that harvesting them produces an amount equivalent to modern fields of domestic wheat. Furthermore, the protein content of wild wheat is twice that of domestic wheat so that one would be hard put to explain how domestication could have arisen in fertile areas. On the other hand, human populations forced to occupy marginal lands have everything to gain from supplementing sparse stands of wild cereals with their domesticated counterparts.

Robert Carneiro (1970) has used a similar model to explain the origin of the state. Carneiro observes that pristine states such as Egypt, Mesopotamia, Harappa, Mesoamerica, and Peru held in common a habitat that was geographically limited; that is, the pristine states developed in coastal valleys, mountain basins, or other areas that restricted movement. The growth of small farming populations in such habitats can initially be handled by movement into uninhabited parts; however, as the size increases, farming land in the habitat is completely filled to its carrying capacity and new increments of growth are left with no place to go. Carneiro's work in the Amazon Basin suggests that if geographical barriers do not prevent movement, simple farming populations are not faced with this problem. When a community reaches a size that is sufficiently large to be no longer supported by the habitat, a segment breaks off, moves away, and a new daughter village is established through the process of "fission." (However, see Cowgill 1975 for a rebuttal to the fission concept.) Furthermore, there is no evidence that warfare in "open" areas of this sort was over land; rather it was conducted for reasons of personal prestige, revenge, and the taking of women. Even if a village were defeated in war, its inhabitants would neither be removed from their land nor incorporated as captives in the social system of the victor.

In a "closed" area, however, the process was completely different. After the available land was filled, intensification of land use began (a la Boserup) resulting in the development of irrigation schemes, terracing, and so forth. However, eventually even this was insufficient to support the rapidly expanding population and squabbles over land became commonplace. Local villages engaged in warfare but for the purpose of taking land, not simply for personal prestige or revenge. Defeated villages could not flee because of the closed habitats and they were either exterminated or, if allowed to stay on their land, forced to give to the victors a portion of the land's produce. Carneiro sees this process as having a snowballing effect. Fights between villages escalated, with increasing population pressure, into fights between regional chiefdoms, then into warfare between populations inhabiting adjacent habitats, and so forth. Eventually, the process triggered the formation of a large state organization.

Although initially Carneiro considered only geographical limitations on land, his most recent statements have added "resource concentration" and "social circumscription" as equivalent to a closed habitat. A resource concentration is a restricted microenvironment containing a valued resource, for example, a mussel shoal, a strip of rich alluvial farmland, or an outcrop of volcanic stone. Social circumscription is based upon the work of Chagnon among the Yanomamo of Venezuela and Brazil referred to several times throughout this book. It is an effect resulting from a high population density, even in open regions. Populations living in the center of the area are prevented from moving outward, not because of geographical barriers but because of the presence of other humans.

The idea of population pressure as a stimulus to social and technological change is combined with the "dismal theorem" of Malthus by the anthropologist Richard G. Wilkinson (1973) to give one of the most interesting theories of economic development in recent years. Wilkinson believes that population growth, as well as environmental change, can upset the stability of the ecological system in which a human population participates. The human population is forced to adapt to the new conditions and responds with a spurt of cultural innovations, mostly technological, to reduce the stress. Population pressure, however, is not relieved by these adaptive changes and *poverty* increases. Increased poverty, in turn, brings about a search for new ways of making a living and stimulates even more cultural innovations. At this point self-sufficiency of the local group is breaking down, imports to cover the deficiency are common, and specialized products are manufactured for export to defray the cost of importing; in short, economic development has set in.

The history of England from the twelfth century on supports the theory. According to Wilkinson "the periods of most acute population pressure—around 1300, 1600 and the late eighteenth century—were all periods of remarkable expansion and innovation as well as of unusual poverty" (1973, pp. 75–76). In the 13th century population growth brought about the cultivation of much new land by clearing forests and draining swamps, among other things. The expansion of farming land was counteracted by a decline in productivity due to shortening of the fallow interval, bringing poor land into production, and not having enough animal manure to sufficiently fertilize both the old and new fields. Liming, ploughing straw ash into the fields, and planting new seed varieties were introduced to improve the yield, but the improvement could not sufficiently reduce population pressure. Thus the price of wheat tripled between the period 1160–1179 and the period 1300–1319. Furthermore, the size of farm holdings was decreasing, reflecting the practice of partible inheritance (farming land is divided among all the sons rather than all of it going to a single son) and continued population growth. Rising prices and less farming land per person increased poverty and the need for sources of income other than farming. The rise of an urban proletariat was one result but more striking was the expansion of industry into the countryside to provide local jobs. Thus mining and smelting of tin, lead and iron; salt production; and pottery manufacture increased greatly and the raising of sheep for wool export jumped 40 percent. Village self-sufficiency began to break down.

A similar spurt of economic development took place from about 1540 to 1640, also on the heels of rapid population growth. Initial attempts to increase food production brought even more land over cultivation "including even the hedgerows which some planted with fruit trees" (Wilkinson, 1973, p. 78). Numerous technological innovations to increase productivity were introduced once again but, as in the thirteenth century, prices soared and poverty ran rampant. Industrialization to provide an income from nonfarming jobs expanded dramatically for the second time in the millenium. Brassmaking was introduced, and steelmaking, glassmaking, salt-boiling, brewing, and brickmaking thrived. The demand for fuel as a source of heat for these industries forced a change from timber to coal, with all the implications for mining and manufacturing that went along with it. Wool exporting was converted into the finished cloth industry to provide even more jobs. By this time local villages were well on their way to being integrated into the vast trading and market networks common to developed nations. A final population spurt in the eighteenth century brought about the "industrial revolution" so well known to us all. It would better be called the industrial "evolution," since it merely continued trends started much earlier.

While there is much to suggest that population pressure is an important factor in culture change, its primacy as a cause cannot be accepted without reservation. In the first place we have not found a suitable method of measurement, and none will be found until the concept of carrying capacity is put on firmer ground. In the second place an ecological perspective forces us to look for *complex* causes not simple ones. Population pressure is immersed in a causal network and cannot be treated as an independent or dependent variable. Thus, in discussing the relationship between population growth and change in subsistence pattern, Don E. Dumond argues for *interdependency:* "a limit on the expansion of subsistence, for whatever reason, means a curtailment of growth in population; a limit on population growth, for whatever cause, entails a lack of expansion of subsistence means" (1972, p. 291).

SUMMARY

In this chapter the concept of carrying capacity has been viewed as important to models of population growth but as simplistic and vague as presently conceived. A final issue must be mentioned in passing. The meaning of carrying capacity as presented here is the *maximum* size than an environment can support. Although this is the meaning usually intended by anthropologists, it is only one of several that have been identified. Another important meaning is the *optimum* size that an environment can support, that is, the size at which resources, living space, and so forth are sufficient to assure the maximum wellbeing of the population's members. Just what this level should be is one of the most hotly debated issues in human ecology today. For example, no one is sure how many calories or grams of protein are needed, and at which intervals, to make sure that an individual reaches full growth and longevity. Thus some recent studies have shown that inadequate protein in the diet of a young child can hinder full intellectual development even though the individual still survives to adulthood and reproduces normally (e.g., Cravioto, Delicardie, and Birch, 1966).

The optimal size of a local group is another case in point. Some have contended that man is tied by biological inheritence to his primeval past and that, therefore, the small families and few social interactions observed in historic hunters and gatherers are optimal for human wellbeing (e.g., Iltis, Loucks, and Andrews, 1970). This view suggests that local groups should be no more than around 500 persons (the size of a hunter and gatherer's usual sphere of interaction). Although there is much to be said for reducing the size of the group with which an individual usually comes into contact, small groups have their own disadvantages. If you have ever lived in a small town, you are already aware of the social devices such as ostracism and rumors that discourage individual idiosyncracies and encourage uniformity. How much might be lost to the population in creative innovation? Until an adequate measure of well-being is available, the concept of optimal carrying capacity is hopelessly bogged down in confusion.

PART THREE
SPECIAL TOPICS IN ECOLOGICAL ANTHROPOLOGY

TWELVE ETHNOECOLOGY

Catherine S. Fowler

In recent years, certain ethnographers have attempted to focus the attention of anthropology in general and human ecology in particular on native views of the environment. Their claim, according to Charles Frake, one of the leaders in the attempt, is that "an ethnographer . . . cannot be satisfied with a mere cataloguing of the components of a cultural ecosystem according to the categories of Western science. He must also describe the environment as the people themselves construe it according to the categories of their ethnoscience" (Frake, 1962a, p.55). Only in this way, continues Frake, can an ethnographer hope to determine "the extent to which ecological considerations, in contrast, say, to sociological ones, enter into a person's decision of what to do" (Frake, 1962a, p. 55).

These statements and others that have appeared since the mid 1950s form the basis of what is generally referred to as *ethnoecology,*[1] a "distinctive approach to human ecology that derives its goals and methods from ethnoscience" (A. Johnson, 1974, p. 87). In this chapter, the history and general methodology (and criticisms of this approach) will be reviewed as well as some of the substantive findings.

HISTORY AND CONCEPTS

Ethnoecology, as an approach to human ecology, was first proposed by Harold Conklin and Charles Frake in a series of thought-provoking

[1]The term ethnoecology is used explicitly by Conklin (1954, 1967). Frake (1962a) refers to essentially the same approach as "ethnographic ecology." Rappaport (1963) proposes a similar concept in his "cognized environment."

papers published from the mid 1950s through the early 1960s.[2] In these papers, Conklin and Frake suggested that ecologically oriented ethnographers would do well to combine traditional techniques from cultural and biological ecology with others designed to more systematically explore native conceptions of their environments. As Frake (1962a, p. 54) notes, "no ethnographer describes social relations in an alien society by referring to the doings of 'uncles,' 'aunts,' and 'cousins.' Many ethnographers do, however, describe the pots and pans, the trees and shrubs and soils and rocks of a culture's environment solely in terms of categories injected from the investigator's culture." Determining native conceptions of these features should be part of any thorough-going ethnographic study. Furthermore, if we are ever to fully comprehend human motives for responding to the environment in particular ways, we shall have to rid ourselves of a priori biases and attempt to learn what the natives "consider worth attending to when making decisions on how to behave within their ecosystem" (Frake, 1962a, p. 55).

Ethnoecology as a method and an orientation to ethnographic ecology is part of the larger and more inclusive field of *ethnoscience.* Ethnoscience, "the new ethnography," or "semantic ethnography" as it is variously known, is roughly the same age and shares many of the same goals, methods and personnel.[3] It employs techniques and concepts developed in descriptive linguistics, systematic biology, and psychology to the exploration of systems of knowledge and cognition in other cultures (Conklin, 1962; Frake, 1962b; Goodenough, 1957; Sturtevant, 1964). The term *ethnoscience* (as well as *ethnoecology*) is in itself a compound. According to William Sturtevant (1964, p. 99), one of the principal reviewers of the goals and methods of ethnoscience, the prefix *ethno-* takes on a new meaning in this context. Whereas the earlier and more typical meaning of the form in such compounds as "ethnohistory," "ethnobotany," and "ethnogeography" was the general study of the history, botany, or geography of an "ethnic group," in this context it refers to any such study done from that group's *own point of view.* Thus, an ethnohistory with an ethnoscientific orientation would be a group's own conception and interpretation of past events; an ethnobotany would be a group's view of the nature of plants and the plant world; and an ethnogeography would be a group's understanding of its location and orientation within geographic space. Ethnoecology would similarly be a group's conception of

[2]See especially Conklin (1954, 1957) and Frake (1962a) although Conklin (1955, 1962, 1967) and Frake (1961, 1962b) also are pertinent.

[3]Contributions by Conklin and Frake are pertinent to the methodology of ethnoscience, particularly Conklin (1955, 1962) and Frake (1961, 1962b).

biotic interrelationships in its universe. Ethnoscience would become the sum total of a group's knowledge, conceptions and classifications of objects, activities and events of its social and material world—the sum total of the

" 'reduction of chaos' achieved by a particular culture." (Sturtevant, 1864, p. 100)

Although those who use the ethnoecological approach tend to agree that the emphasis of the method is in keeping with this definition of *ethno-*, they also in practice include materials and approaches that are more appropriate to the *ethnic group* or *cultural ecology* of previous fieldworkers.

This difference in the ethnoscientific orientation follows by analogy one that has been part of the conceptual framework of linguistics for many years: namely, the difference between *etics* and *emics.*[4] A linguist beginning to record and describe the sound system of a language commonly uses a system of phon*etic* transcription. Such a system, as for example, the International Phonetic Alphabet (IPA) or some modification of it, provides the linguist with a set of detailed, yet language-free sound symbols from which he or she may choose those approximating the sounds of the language under consideration. The transcription of any particular language will not require all of the symbols. In fact, since IPA and other such systems were devised from the combined experience of linguists working with languages with many types of sound systems, it contains far more than would be necessary. Such systems are designed to be universally applicable.

Once a language has been recorded phonetically, the linguist will proceed to a phon*emic* analysis. At this level, he or she will examine the sound system of the language in question *from its own perspective.* Details and factors that can be accounted for by conditioning features in the environment of the sounds[5] will be sorted out and discussed. Only those sounds that are language-specific, ór peculiar to the structural system of sound in that particular language, will remain as its phonemes. They will more closely approximate what is *significant* to the sound system of that

[4]The differentiation in linguistics, however, remains as that between phon*emics* and phon*etics* although other levels of linguistic analysis employ similar principles; for example, morpheme to allomorph. See also Chomsky (1965) for the "deep structure" versus "surface structure" distinction, which is also pertinent.

[5]The environment of a sound includes the sounds adjacent to it. They will cause a shift in articulation that will be recorded phonetically; e.g., the difference in "k" sounds next to the front, mid, and back vowels [i], [a], and [u] as in English key, caw, and coo.

language than will the overparticularized phonetic transcription. These sounds are those that are uniquely relevant for the speakers—and, as Edward Sapir (1933) once suggested, those that have a peculiar psychological reality.[6]

The relationship of phonetics to phonemics in language is essentially the same as that of *etics* to *emics* in the analysis of culture. The terminology was first suggested by Kenneth Pike in a work titled *Language in Relation to a Unified Theory of the Structure of Human Behavior* (Pike, 1954–1960). For Pike, cultural etics are the culture-free features of the social and material world. If not truly culture-free, they will be those that have been derived from the examination of many cultures, or the sum of all significant attributes in the folk classifications of all cultures (Sturtevant, 1964, p. 102). They are the ethnographer's stock and trade. They allow categorization of cultural activities into spheres such as social organization, political organization, technology, ideology, and many more. A knowledge of the total cultural variation possible allows the ethnographer to begin to describe the particular culture under scrutiny. Cultural etics allow the ethnographer to bring order out of the seeming chaos of cultural behavior (Sturtevant, 1964, p. 102).

Cultural emics thus became the culture-specific aspects of a people's world. The emic approach also becomes essentially the same as an ethnoscientific one (Sturtevant, 1964, p. 102). It more closely approximates the analysis of culture from its own perspective. Details that can be accounted for by various conditioning features may be sorted out and discussed. Only those aspects that are locally significant will remain. These should have a unique relevance to the bearers and be "cognitively salient" (Wallace and Atkins, 1960, p. 64). In other words, they should more closely approximate the cognitive structure of the culture—its psychological reality.[7]

Thus far, research within ethnoscience and ethnoecology using these principles has concentrated on native systems for classifying the natural and social phenomena of their worlds (Sturtevant, 1964, p. 105). Included are studies of kinship terminologies, color categories, residence rules, disease terminologies, ethnobotanical and ethnozoological systems, and many more (see Sturtevant, 1964; Colby, 1966; Hammel, 1965; and Conklin, 1972 for bibliography). Those of particular interest to ethnoecology include many of the same studies in ethnobiological classifications, others

[6]Psychological reality is not the principle means by which phonemes are defined, however. Distributional and other factors characterize phonemic methods.

[7]On this latter point there is considerable discussion, to which we shall return later (see "Critique and Discussion").

dealing with various aspects of subsistence (fishing, agriculture), settlement patterns, and so forth. Studies dealing with native classifications have been appropriately labeled "studies in ethnosystematics" by Andrew Vayda and Roy Rappaport (1968, p. 491).

Classificatory systems, particularly as reflected in native terminology, are felt by many to hold important clues to understanding the mental processes by which cultural knowledge is coded and recalled (Conklin, 1962; Frake, 1962b). From the analysis of native classification schemes, investigators hope to be able to extract "the rules by which a people decide on the category membership of objects in their experience" (Frake, 1962a, p. 55). From these rules, they hope to proceed to more complex ones that may predict other aspects of behavior. In the context of ethnoecology, these might be rules that determine the killing of game, the clearing of fields, the building of houses, the changing of residences, etc. (Frake, 1962a, p. 55). According to Frake,

The analysis of a culture's terminological systems will not, of course exhaustively reveal the cognitive world of its members, but it certainly will tap a central portion of it. Culturally significant cognitive features must be communicable between persons in one of the standard symbolic systems of a culture. A major share of these features will undoubtedly be codable in a society's most flexible and productive communication device, its language. (1962b, p. 75).

The features of language commonly chosen for analysis in native classifications are *lexemes*. A lexeme, as defined by Conklin (1962, p. 121) is "a meaningful form whose signification cannot be inferred from a knowledge of anything else in the language." Lexemes may be morphemes, words, complexes, phrases, or other units of speech that serve to label *segregates*, or "any terminologically-distinguished . . . grouping of objects" (Conklin, 1962, pp. 120–121). In many cases, the referent of a lexeme will be a segregate that is unrelated to the typical referents of its component linguistic parts. Thus, for example, a 'hot dog' is hardly a 'sizzling canine', nor is a 'jack-in-the-pulpit' a 'preacher named Jack.' (see Conklin 1962, for other examples). Since linguistic expressions of various structural types are often prominent features of native classification schemes, and since linguistic form and meaning need not be isomorphic, a concept like *lexeme* is required in the methodology of ethnoscience (Conklin, 1962).

Segregates that are related so as to form a distinct sphere of meaning, a major category, or a classification system within a given culture constitute a *domain*. Domains are culture-specific. In other words, their boundaries and contents must be discovered and defined in terms of the culture being investigated rather than in terms of some external or cross-cultural criterion (Sturtevant, 1964, p. 104). What may constitute a single domain for one culture may not be a separate sphere for another. For example, although for speakers of English, 'plants' usually constitute a single domain capable of being classified in various ways, for speakers of Northern Paiute, a North American Indian language, this is less clearly so. Rather, they are part of a larger and interconnected domain of foods, medicines and so forth. Nor is there a single term in Northern Paiute equivalent to English "plant" (Fowler and Leland, 1967). Determining the boundaries, contents, and external and internal relationships for domains in different languages and cultures becomes the primary objective of the methodology of ethnosystematics in ethnoscience or ethnoecology.

METHODS OF ANALYSIS

In order to begin an investigation into a native terminological system or classification scheme, a corpus or body of data is first required. These data will usually consist of native terms, such as for example, the names for plants, animals, insects, or various aspects of the environment such as its geographical features, etc. These may have been gathered by traditional ethnographic procedures,[8] or through one of the specific techniques of ethnosystematics to be discussed below. All of the terms should relate to a single domain, thereby sharing at least one feature of meaning in common (Conklin, 1962; Frake, 1962b).

Within this context, the attempt is then made to relate each term to the others in its set. The approach chosen will partly depend on the nature of the terms and the relationships among them. The most common relational systems for native classifications are, however, *taxonomies, paradigms,* and *keys.*[9]

Taxonomies

Terms related taxonomically are those that show hierarchical relationships (Conklin, 1962; Frake, 1961, 1962b). Taxonomies are perhaps most familiar to us in the form of botanical and zoological classifications.

[8]A frequent task for all ethnographers is to ask the "names" of things, or to attempt to focus discussions on key native concepts.

[9]See particularly Tyler (1969) for a useful summary of methods. Sturtevant (1964) also reviews componential analysis and taxonomies. Kay (1969) discusses other types of semantic relationships not treated here.

In these, there is the general-to-specific arrangement from *kingdom,* to *phylum,* to *class,* to *order,* to *family,* to *genus,* to *species,* to *subspecies* or *variety*. Taxonomies of other domains would contain other divisions, but the principles of arrangement would remain roughly the same (Conklin 1962).[10] Taxonomies are commonly presented in the form of multibranching tree diagrams (see Figure 12.1 for partial example).

In taxonomies, terms and, by implication, their referents (objects, activities, etc.) are related to each other according to the principles of *inclusion* and *contrast* (Frake, 1962b). Terms related by *inclusion* are those that can be hierarchically arranged from general to specific; for example, plant > tree > conifer > pine > ponderosa pine. Such forms are often discussed in the context 'X is a *kind of* Y,' 'ponderosa is a *kind of* pine,' 'pines are *kinds of* conifers.' Terms related by *contrast* are those presented on any single horizontal dimension of a taxonomy or tree. They are held to be mutually exclusive in meaning. For example, 'pinyon pine' contrasts with 'ponderosa pine' and 'pine' with 'fir' (see Figure 12.1). In the context of a discussion each would commonly be kept distinct, such as in the sentence, 'Look at that (ponderosa pine, pinyon pine, *or* pine, fir),' but *not* 'Look at that ponderosa pinyon pine' or 'Look at that pine fir.'

Several suggestions for eliciting native taxonomies have been made by those working in ethnosystematics. Each technique attempts to shift

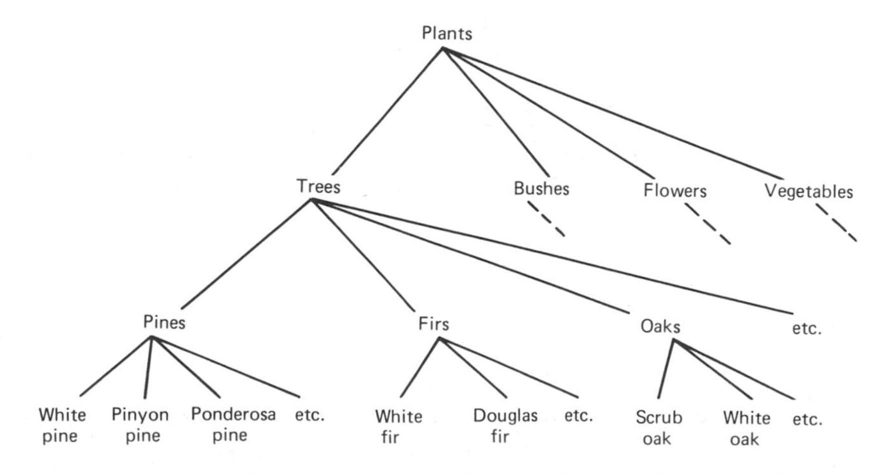

FIGURE 12.1 Hypothetical taxonomy of part of the plant domain in American English.

[10]See, however, Bright and Bright (1965) for mutually influencing taxonomies.

control of the eliciting situation to the native informant, in the hope that more culturally relevant data will be forthcoming. Perchonock and Werner (1969) provide a useful review of several of these techniques.

DRAWING TREE DIAGRAMS. Native informants are asked to draw taxonomic tree diagrams in any manner they feel best expresses the relationships in question. They are also asked to label with a native term each appropriate node or intersection in the tree.

SLIP SORTING. Informants are asked to sort cards or slips of paper, each containing a native term, or perhaps a picture of the item in question, into various piles according to principles of their choosing. Piles may be separated or combined as new principles come to mind. Informants are to supply a native term for the appropriate piles.

DIRECT QUESTION/RESPONSE. Informants are questioned directly about taxonomies through a controlled set of question frames. Appropriate responses to each question then become the focus of a new frame until a terminal response is reached. Table 12.1 illustrates one form of this technique, that suggested by Metzger and Williams (1966), as applied to data on Northern Paiute ethnobiology.

ELICITING FOLK DEFINITIONS. Each native term becomes the focus of a folk definition, with the definition subsequently being analyzed for taxonomic elements as well as other relational principles.

KEY WORD IN CONTEXT. Larger native language texts are analyzed for taxonomic and other words and relationships. These are systematized and additional questions are asked about them.

NATIVES AS ELICITERS. Native informants are employed as elicitors and directed to ask questions they feel pertinent, or they are asked to supply the investigator with pertinent questions so that both can participate in the discussion. Contextual discussions are then analyzed for taxonomic materials and for rules of usage.

Each of these techniques can be and has been useful to certain investigators in certain circumstances. However, none is without at least some constraints and biases such that additional techniques may be required (see Fowler and Leland, 1967; Perchonock and Werner, 1969). Some of these will be discussed in the review of examples of specific studies in ethnosystematics that follows this section.

TABLE 12.1 NORTHERN PAIUTE TAXONOMIES: DATA DERIVED BY APPLICATION OF QUESTION-FRAME ANALYSIS TECHNIQUE[a]

Q.	[híi tuʔíhi kadíipɨgubagweitɨ]	
	What is everything on or above the earth?	
A.	[tɨdɨkásan·a]	*Things we eat*
Q.	[him·ádɨwazu]	*What else?*
A.	[tɨdɨhóaweisan·a]	*Things we hunt*
Q.	[him·ádɨwazu]	*What else?*
A.	[padúhatɨ]	*Things under the water*
Q.	[him·ádɨwazu]	*What else?*
A.	[yozɨ́dɨ]	*Things that fly*
Q.	[him·ádɨwazu]	*What else?*
A.	[óʔnosaba]	*That's all*
Q.	[híi tɨkásan·a]	*What is eaten?*
A.	[akɨ́, tɨbá, kuhá, acá, kam·ɨ́,] . . .	
	Sunflower, pine nut, blazing star, tansy mustard, rabbits, . . .	
Q.	[híi tɨhóaweisan·a]	*What is hunted?*
A.	[tɨhɨ́ʔya, kóipa, tɨn·á]	*Deer, mountain sheep, antelope*
Q.	[híi yozɨ́dɨ]	*What flies?*
A.	[huzíba, pɨhɨ́, páanosa, wohítya]	
	Birds, duck, pelican, swan	
Q.	[híi padúhatɨ]	*What is under the water*
A.	[agái, kuyúi]	*Trout, cui-ui*
Q.	[him·á sunɨ́mɨ kái tɨkásan·a]	*What don't the Indians eat?*
A.	[togóg·wa, mugúzu, pamógo, soáda, pipúzi, tɨb·óca, kagwíduhuʔu	
	Rattlesnake, lizard, frog, spider, stink bug, lizard, mountain lion	

Source. Catherine S. Fowler and Joy Leland, "Some Northern Paiute Native Categories," *Ethnology*, 6, 1967. Copyright © 1967 by the University of Pittsburgh Press. Reprinted by permission of the publisher.)

[a] *Metzger and Williams, 1966.*

Taxonomies do not tell us the absolute meaning of the terms in question; they merely relate them from general to specific. In addition, there are also domains, or in some cases sections of domains, that are not organized taxonomically. In order to define native terms more precisely as well as deal with other forms of relationships, two additional techniques are often applied. These attempt to relate more systematically sets of semantic features. They are known as *paradigms* and *keys* (Tyler, 1969).

Paradigms

Paradigms define terms on the basis of intersections of features. The notion of paradigm can be illustrated most simply by a block diagram such as that provided in Figure 12.2. In this diagram, characteristics *A* and

	A	*B*
C	*AC*	*BC*
D	*AD*	*CD*

FIGURE 12.2 Paradigm of arbitrary features *A, B, C,* and *D.*

		Sex		
		Male ♂	*Female* ♀	*Neuter* ϕ
Maturity	Adult M–1	Stallion Boar	Mare Sow	Gelding Barrow
	Adolescent M–2		Filly Gilt	
	Child M–3	Colt Shoat		
	Baby M–4	Foal Piglet		

FIGURE 12.3 Paradigm of features for "horse" and "swine." (From Stephen Tyler, "Introduction," in Stephen Tyler, ed., *Cognitive Anthropology,* New York: Holt, Rinehart, and Winston, 1969. Copyright © by Holt, Rinehart, and Winston, Inc. Reprinted by permission of the publisher.)

B intersect with characteristics *C* and *D* to produce *AC, BC, AD,* and *BD.* In semantic paradigms, these "characteristics" are individual components or features of meaning and the technique for discovering them is referred to as *componential analysis* (Goodenough, 1956). These components can be viewed as intersecting to produce the terms in question and at the same time, their definitions. Figure 12.3, drawn from Stephen Tyler's (1969, p. 10) analysis of certain terms for livestock in American English, illustrates the application of this principle. Each term in this set lies at a certain point in the intersection of the semantic components "maturity" and "sex." Those points are further defined by the subcategories "adult," "adolescent," "child," and "baby," and "male," "female" and "neuter," respectively. The term 'stallion' can thus be defined paradigmatically as "adult, male" of the "horse" grouping and comparable to 'boar,' the "adult, male" of the "swine" grouping (Tyler, 1969, p. 10). Structural definitions such as these, as well as the display of these materials in block diagrams, often allow the investigator to see more clearly the systematic aspects of dimensions of meaning (Goodenough, 1956). Discovery of the relevant semantic components that differentiate members of a set or domain is not always easy, however (Tyler, 1969). Paradigms of varying types are used frequently in linguistics to describe features of the sound systems and grammar as well as features of meaning.[11]

Keys

In addition to paradigms, semantic features may also be arranged according to a *key*. Keys contrast single features of meaning rather than items or terms (Tyler, 1969). They are based on a series of choices between alternative attributes, usually designated as present (+) or absent (−). Keys are perhaps most familiar in the context of botanical identification, wherein the particular plant specimen in question is examined for the presence or absence of a number of trait attributes (see Figure 12.4). A positive or negative choice made at any one juncture in a botanical key will lead the identifier along one path only. Ultimately, when all appropriate choices have been made, the specimen is identified. Definitions produced through the use of keys would take on the character of attribute lists. They could be understood only in the context of other terms so defined. Thus, in the accompanying figure (Figure 12.4) illustrating the analysis of a small segment of the genus *Pinus* (pines), a 'foxtail pine' *(Pinus balfouriana)* has needles in clusters of five (+), and cone scales that are thick (+) and prickled (+) with the prickles being incurved (+). If the latter feature is lacking, that is, prickles *not* incurved (−), the form is a 'bristlecone pine' *(P. aristata)*.

Taxonomies, paradigms, and keys are the primary types of relational systems thus far noted in native classification schemes. However, they are not the only types possible (see Kay, 1969). Of the three, taxonomies appear to be the most common, at least in the literature (see Conklin, 1972). Examples of paradigmatic schemes are less common with domains

[11]The definition of sounds by point of articulation and manner of articulation is an example. Case and tense conjugations for nouns and verbs are also paradigms.

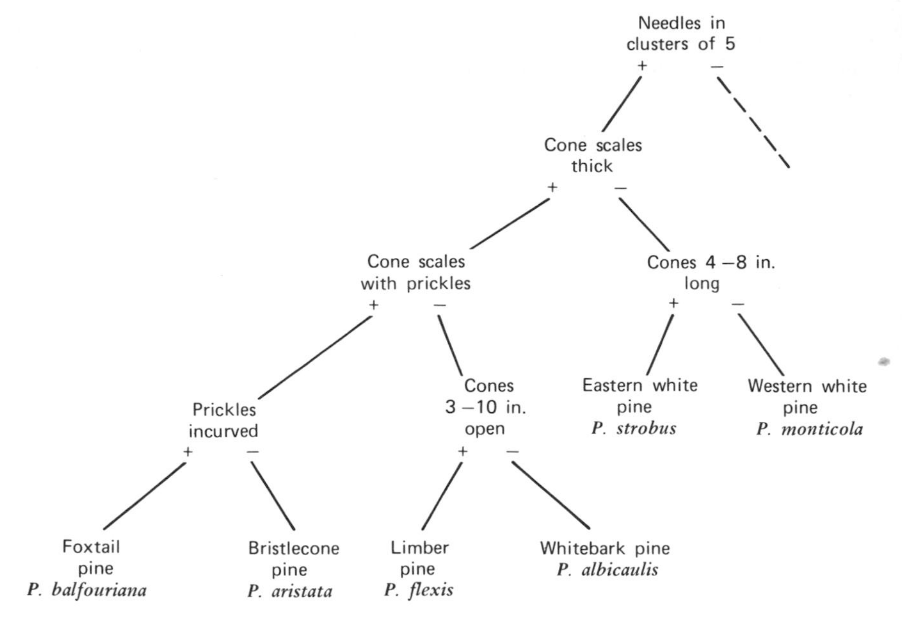

FIGURE 12.4 Partial key to species of pines (*pinus*). (Data from Preston 1966, p. 3.)

described by keys having the lowest frequency of all. The high frequency of taxonomic schemes may be due in part to the nature of the domains investigated thus far (i.e., ethnobotany, ethnozoology, and so on), in part to the relative ease with which the technique may be applied in the field, but perhaps also in part to the general recurrence of this principle of ordering. Frake (1962b, p. 81) at least, has suggested that "the use of taxonomic systems . . . is a fundamental principle of human thinking."

Some specific examples of ethnoecologically oriented studies of native classifications, or what Vayda and Rappaport (1968, p. 491) term "studies in ethnosystematics," are discussed below. The examples have been chosen to illustrate the techniques discussed as well as to raise some methodological questions. Following these, some studies will be presented that deal with other aspects of ethnoecology, concluding with a general discussion and critique of the approach.

STUDIES IN ETHNOSYSTEMATICS

The fields of ethnobotany and ethnozoology provide ample possibilities for demonstrating the principle of taxonomy in native terminological systems (see Conklin, 1972, Sections 4 and 5). However, studies in these areas in the past rarely probed the native ordering of any of these various domains in an exhaustive way. Until the introduction of the methods of ethnoscience, many works in these areas were of the "economic" type: that is, they were catalogs of the *uses* of plants and animals by native peoples.[12] Even less frequently was the *total* biotic universe of a people investigated; in other words, the plants and animals *not* utilized in the local environment as well as those utilized. These and other problems make many of the older works difficult to interpret from the point of view of the correlation of native taxonomic concepts with those of Western scientific botany. A fresh approach to Frake's (1962a, p. 54) "pots and pans, . . . trees and shrubs and soils and rocks of a culture's environment" was thus needed.

To date, the most exhaustive taxonomic study available in ethnobotany is that by Brent Berlin, Dennis Breedlove, and Peter Raven (1974) titled *Principles of Tzeltal Plant Classification: An Introduction to the Botanical Ethnography of a Mayan-Speaking People in Highland Chiapas.* In this work, the authors—an anthropologist and two botanists—examine the taxonomic relationships of 471 natively defined generic taxa.[13] These forms are defined as generics (i.e., comparable for the most part to the *genera* of Western systematic botany) based on certain linguistic features of their names as well as on their taxonomic function. They are generally the terminal, or lowest-level responses in the taxonomic scheme, and they are also single units semantically; that is they label distinct plant entities or segregates. The Tzeltal do not have a single term comparable to "plant" in English, the highest level of generalization. However, it is clear to the authors that plants form a domain conceptually as they are so treated in discussions. Plant names are also marked linguistically by the syntactic requirement of a specific numeral classifier /tehk/ in discussions involving the counting of plants as opposed to animals or other objects (Berlin, Breedlove, and Raven, 1974, p. 30).

In Berlin, Breedlove and Raven's work, each Tzeltal generic term is associated with as many plant genera and species identified according to the terminology of Western systematic botany as the Tzeltal feel appropriate. This includes those that are its most common referents (referred to as its *basic range*) as well as those that might be included as the extended

[12]The works of Malkin (1956, 1958, 1962), Robbins, Harrington, and Freire-Marreco (1916), and Wyman and Bailey (1964) are examples of early exceptions.

[13]Taxa, according to Berlin, Breedlove, and Raven (1974, p. 25) are "linguistically recognized groupings of organisms of varying degrees of inclusiveness;" e.g., *oak, vine, plant, red-headed woodpecker* in English.

referents *(extended range)* of its meaning. Since the authors made a thorough inventory of the locally occurring plant species (15,000 specimens were collected in one region alone) they feel that the data provided on the basic and extended ranges for the terms are probably close to complete (Berlin, Breedlove, and Raven, 1974, p. 153). The correlation of Tzeltal generics with those of Western scientific botanists is thus accomplished. The authors also include the numbers of informants and numbers of specimens involved in the identifications (see Table 12.2, numbers in parens are informants and specimens, respectively).

The Tzeltal genera are organized into certain higher order relationships based on Tzeltal perceptions of features of plant gross morphology. The taxonomic levels are /teʔ/ 'trees,' /wama/ 'herbs,' /ʔak'/ 'vines,' and /ʔak/ 'grasses.' These four categories, termed "life form" categories by the authors, group and account for about 75 percent of the Tzeltal generics. The remainder are recognized as unaffiliated and/or ambiguous generics; that is, they do not fit into classes, or could be in more than one class (Berlin, Breedlove, and Raven, 1974, p. 30). Each of these major classes is further subdivided into a series of midlevel classes that relate various generics into complexes. Nearly all of these complexes are unlabeled linguistically. However, they constitute covert categories of relationship based on informants' judgments. Figure 12.5 presents an example of two Tzeltal complexes within the unit /teʔ/ 'trees.' Some complexes are multibranching, containing several degrees of covert relationships. The recognition that taxonomic relationships may be covert and may lack linguistic labels is an important methodological and theoretical point in the study of taxonomies generally. It is particularly significant as a cautionary note to those who would depend too heavily on linguistic clues and data for evidences of relationships (Berlin, Breedlove, and Raven, 1974, p. 36).

TABLE 12.2 TZELTAL ETHNOBOTANY: BASIC AND EXTENDED RANGES FOR /SIBAN/ COMPLEX

Term	Basic Range	Extended Range
1. *siban*		
bac'il siban	*Cornus excelsa* (11–14)	*Philadelphus mexicanus* (4–3)
cahal siban	*Acalyphya unibracteata* (3–3)	*Acalypha* cf. *vagans* (1–2)
	A. cf. *mollis* (1–3)	
2. *sak hi*	*Cornus disciflora* (9–12)	*Hampea longipes* (2–2)
		Eupatorium ligustrinum (2–4)

Source. Brent Berlin, Dennis Breedlove, and Peter Raven, *Principles of Tzeltal Plant Classification*, New York: Academic Press, 1974. Copyright © 1974 by Academic Press, Inc. Reprinted by permission of the publisher.

Covert Complexes: *teʔ*

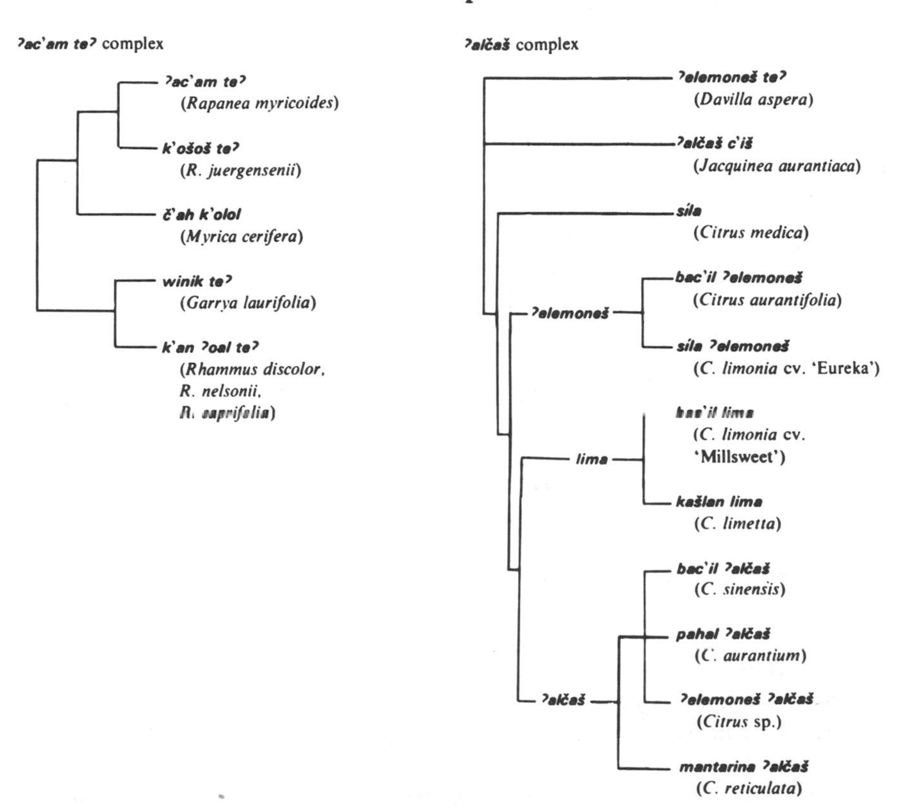

FIGURE 12.5 Tzeltal covert complexes: te (trees). (From Brent Berlin, Dennis Breedlove, and Peter Raven, *Principles of Tzeltal Plant Classification,* New York: Academic Press, 1974. Copyright © 1974 by Academic Press, Inc. Reprinted by permission of the author and publisher.)

Below the level of the genus, the Tzeltal also partition a number of forms into divisions comparable to Western botanical species. Most of these are binomials linguistically, as are those of Western botany (e.g., pinyon pine and white oak). Tzeltal specifics are most frequently generics partitioned according to size, color, shape, texture, habitat, and so forth (see Table 12.3 for examples).

Given the exhaustiveness of their study, Berlin, Breedlove, and Raven are able to attack the problem of the correspondence of Tzeltal and Western botanical classification with a degree of confidence. Specifically,

TABLE 12.3 EXAMPLES OF TZELTAL SPECIFIC NAMES

Generic Names	Specific Names
ˀahateˀ 'white sapote' *(Casimiroa edulis)*	*sakil ˀahateˀ* 'white white sapote'
	k'anal ˀahateˀ 'yellow white sapote'
	celum ˀahateˀ 'elongated white sapote'
sc'ul 'amaranth' *(Amaranthus* spp.)	*sakil sc'ul* 'white amaranth' *(A. hybridus)*
	cahal sc'ul 'red amaranth' *(A. cruentus)*
	ĉ'is sc'ul 'spiny amaranth' *(A. spinosus)*
pehtak 'prickly pear' *(Opuntia* sp.)	*sakil pehtak* 'white prickly pear'
	cahal pehtak 'red prickly pear'

Source. From Brent Berlin, Dennis Breedlove, and Peter Raven, *Principles of Tzeltal Plant Classification*, New York: Academic Press, 1974.

they are able to map with some assurance the distribution of Tzeltal generics with reference to their Western counterparts. In so doing, they suggest the following types of relationships:

1. *One-to-one correspondences,* or cases where a single Tzeltal generic refers to one and only one botanical species (this obtains for 291 of the 471 Tzeltal generics, or roughly 61 percent).

2. *Underdifferentiated,* of two types,
   a. Those in which a single Tzeltal generic refers to two or more species of the same botanical genus.
   b. Where a single Tzeltal generic refers to two or more species of two or more botanical genera (these two types account for 163 of the 471 generics of 35%).

3. *Over-differentiation,* where two or more Tzeltal generics map into a single botanical species (this accounts for only 17 of the 471 generics or approximately 4% of the cases).

They thus conclude a substantial amount of similarity between the two schemes. A number of the cases of nonmatches are explainable on morphological grounds (Berlin, Breedlove, and Raven, 1974, p. 101).

An example of taxonomy for plants based on principles other than morphology is provided by Fowler and Leland (1967) in their sketch of Northern Paiute plant and animal categories. The Northern Paiute are American Indians of the Great Basin of Western North America. In this particular scheme, 'plants' are not the focal domain, but rather form a subordinate one to the major categories, "things that are eaten,' 'things that are used,' and 'things that are not used.' Plants, or more literally, "growers' or 'things that grow in a place,' /naadɨ/, are partitioned among each of these categories and classified differently within each. Under the 'eaten' category, they are differentiated into two classes based on habitat: 'things in the ground' and 'things in the water.' The first is partitioned into the categories 'seeds,' 'berries,' 'greens,' 'roots,' and 'fleshy plants,' each with specifically named members. The second is not further partitioned (see Figure 12.6). Plants in the category 'things that are used' show the same land versus water distinction; however, there are also subcategories within the former for 'medicines,' 'gums,' 'grasses,' 'forest' materials, and so forth. Methods of preparation for medicines further differentiate the plants in this class. Water plants can be listed by name without additional partitioning.

The 'things that are not used' category includes the interesting subcategories 'just flowers,' 'just grass,' 'just willows' (i.e., not used in basketry), 'sticker plants,' and others. An alternative name for this entire grouping of plants is 'just trash' (Fowler and Leland, 1967, p. 393).

The mixed hierarchical character of this classification scheme is obvious. Although the principle of *use* guides partitions at the upper level, considerations of habitat, morphology, use, preparation techniques, by-products, and so forth partition the lower sets. This in part results from the focus of Northern Paiute names on the *products* that plants produce rather than on the plants themselves. In discussions, the terms refer to foods, medicines, and fibers, rather than to the structural plants themselves; for example, /tɨba/ 'pine nut' is the commonly used form; /tɨbapi/ 'pine nut tree' is rarely used. They also note that the major categories ('things that are eaten,' 'used,' 'not used') are roughly ordered in terms of importance with 'things that are eaten' taking precedence. Fowler and Leland (1967, p. 394) also report that the expression *kind of*, as in '*X* is a *kind of Y*' commonly used for eliciting taxonomic relationships is not easily expressed in Northern Paiute. This makes it difficult to apply some of the techniques for eliciting taxonomies in this language.

Although a complete study of the type Berlin, Breedlove, and Raven (1974) produced for Tzeltal plants is not available for a zoological domain

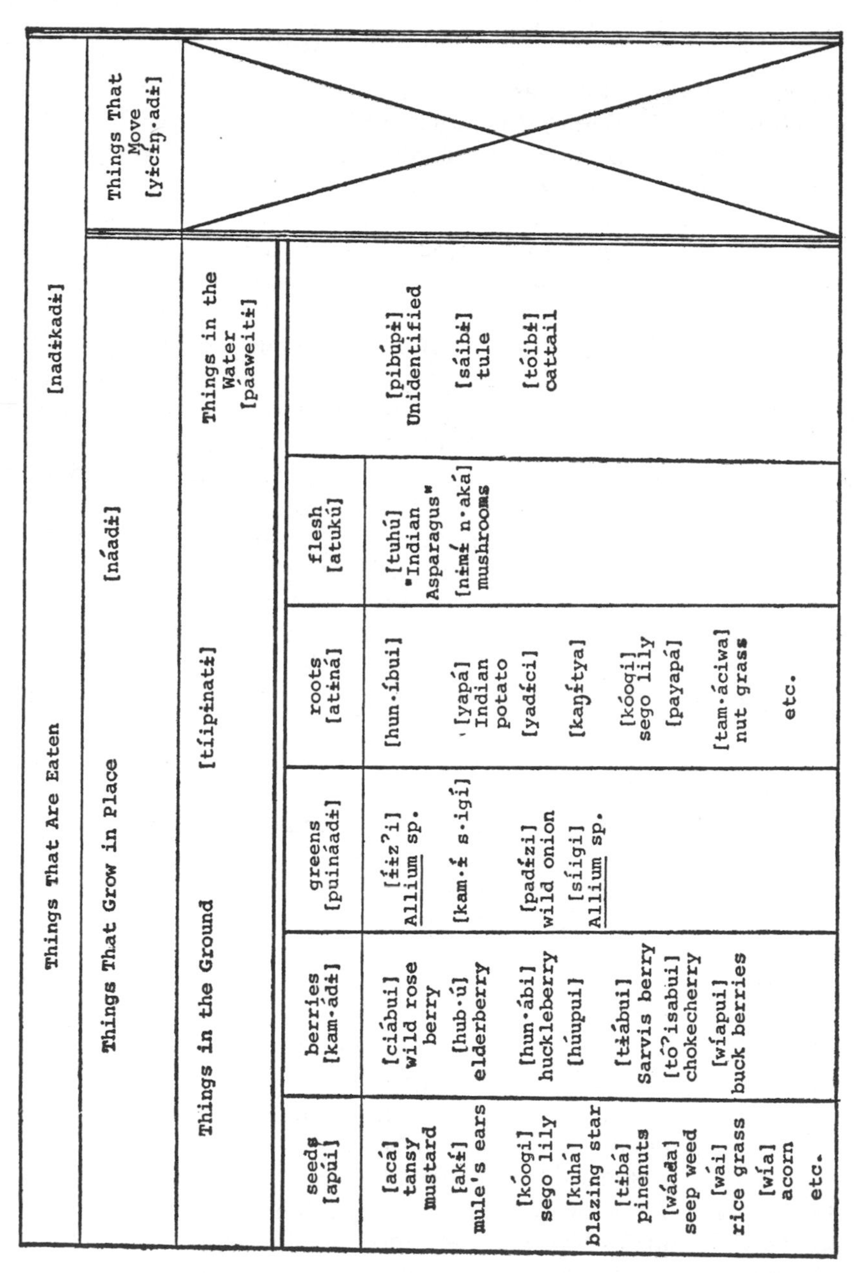

Things That Are Eaten [nadɨkadɨ]						
Things That Grow in Place [náadɨ]						Things That Move [yɨcɨ́ŋ·adɨ]
Things in the Ground [tíipɨnatɨ]					Things in the Water [páaweitɨ]	
seeds [apúi]	berries [kam·ádɨ]	greens [puináadɨ]	roots [atɨná]	flesh [atukú]		
[acá] tansy mustard [akɨ́] mule's ears [kóogi] sego lily [kuhá] blazing star [tɨbá] pinenuts [wáada] seep weed [wái] rice grass [wía] acorn etc.	[ciábui] wild rose berry [hub·ú] elderberry [hun·ábi] huckleberry [húupui] [tɨábui] Sarvis berry [tóʔisabui] chokecherry [wíapui] buck berries	[ɨ́ɨzʔi] Allium sp. [kam·ɨ́ s·igí] [padɨ́zi] wild onion [síigi] Allium sp.	[hun·íbui] [yapá] Indian potato [yadɨ́ci] [kaŋɨ́tya] [kóogi] sego lily [payapá] [tam·áciwa] nut grass etc.	[tuhú] "Indian Asparagus" [nɨmɨ́ n·aká] mushrooms	[pibúpɨ] Unidentified [sáibɨ] tule [tóibɨ] cattail	

FIGURE 12.6 Northern Paiute category "things that are eaten." (From Catherine S. Fowler and Joy Leland, "Some Northern Paiute Native Categories," *Ethnology*, 6, 1967. Copyright © 1967 by University of Pittsburgh Press. Reprinted by permission of the author and publisher.)

in any culture, a number of partial statements relative to aspects of this sphere have been made. One such study in progress will be cited as an additional example of a taxonomy. It also illustrates the use of a *key* as an alternative means of categorization.

Ralph Bulmer has published a number of notes on ethnosystematics in the zoology for the Karam, a horticultural people of New Guinea.[14] Bulmer (1967, p. 6) notes that Karam zoological taxonomies can be approached from two perspectives: One is to examine the general categories or broadest level taxa (of which there are roughly 94), and try to understand the principles of their relationships; the other is to focus on the smallest units that the Karam segregate to see what types of discriminations are made. The latter approach, according to Bulmer (1967, p. 6) indicates that "at this level Karam show an enormous, detailed and on the whole highly accurate knowledge of natural history, and that though, even with vertebrate animals, their terminal taxa only correspond within about 60 per cent of the cases with the species recognized by the scientific zoologists, they are nevertheless in general aware of species differences among larger and more familiar creatures." Bulmer adds that it is his feeling that the Karam classification scheme at this level is dealing with "objective discontinuities in nature" and that there will be "conceptual" correspondences between the great majority of terminal taxa applied by the Karam and the species recognized by the zoologist. In a later publication Bulmer (1970) goes on to discuss this correlation in greater detail, noting that most of the underdifferentiated and overdifferentiated forms in the remaining 40 percent of Karam terminal taxa can be explained as either unfamiliar, not utilized, or species where the biologist also finds it difficult to judge the relationships.

At the upper levels of Karam taxonomies, however, features other than objective biological observations come into play. These are the levels where culture and experience impinge directly, with the result that a number of alternative taxonomic possibilities are available. The correlation to Western classifications, based as they are on morphological, genetic, and other features, is not as good. Bulmer (1967, p. 8) presents one such possibility in the form of a key to Karam segregates. This key, based on general morphological features, can be used to separate 14 of the 20 focal primary taxa for vertebrates and culturally significant invertebrates (see Figure 12.7). He also presents a multidimensional habitat diagram for the same taxa (see Figure 12.8), illustrating an alternative classification by the Karam of these same forms (Bulmer 1967, p. 9). He goes on to suggest that many additional aspects of Karam culture must be considered before

[14]See Bulmer (1957, 1967, 1968a, 1968b, 1970) and Bulmer and Tyler (1968).

the full process by which the Karam decide on the category membership of forms is actually understood. Criteria for decision making may be multiple and complex (Bulmer, 1967, p. 16).

As a specific example illustrating the use of a paradigm and componential analysis in ethnosystematic studies, we shall examine briefly Harold Conklin's (1967) discussion of land-form categories among the Ifugao, an agricultural people of the Philippines. According to Conklin (1967, p. 105), the Ifugao "distinguish hundreds of terrain variations involving combinations of rock, soil, water, vegetation, and the results of

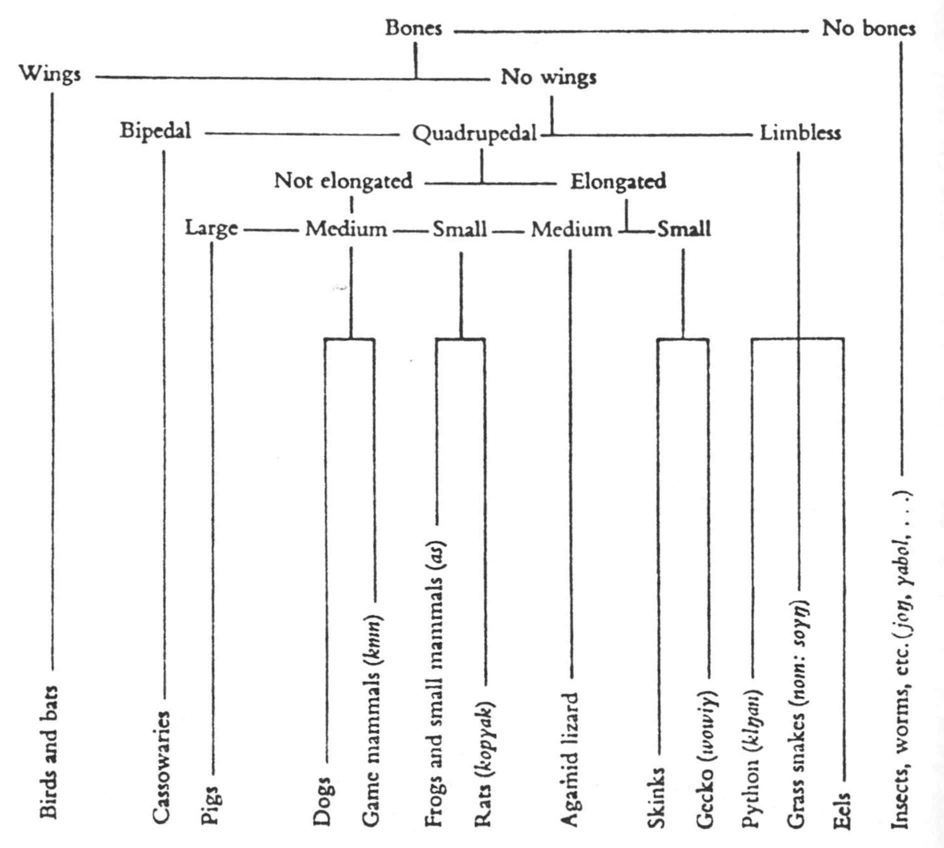

FIGURE 12.7 Key to Karaem primary taxa by gross morphological features. (From Ralph Bulmer, "Why Is the Cassowary Not a Bird? A Problem of Zoological Taxonomy Among the Karam of the new Guinea Highlands," *Man*, (n.s.), 2, 1967. Reprinted by permission of the Royal Anthropological Institute of Great Britain and Ireland.

	Horizontal axis*				Vertical axis†
	Forest	Open country	Gardens	Homesteads	
Birds and bats	———	———	———	———	A, B, C, some Q, some T.
Cassowaries	———				T.
Pigs		- - -	———	———	T.
Dogs	———	- - -	- - -	———	T.
Game mammals	———	- - -			Mainly B or T, some S.
Frogs and small mammals	———	———	- - -		C, Q, T, S.
Rats			———	———	T, S.
Agamid lizard		- - -			B, but lays eggs in ground.
Skinks	———	———	———	———	B, C, T, S.
Gecko			———	———	B, but especially found in bana palms and old houses.
Python	- - -	- - -	- - -		B.
Grass snakes		———			T, S, sometimes Q.
Eels	- - -	- - -	- - -		Q, rarely T.

*Continuous line indicates characteristic habitat; dotted line indicates restricted occurrence.
†A = aerial; B = arboreal; C = found in low vegetation; Q = aquatic; T = terrestrial; and S = subterranean.

FIGURE 12.8 Classification of Karaem primary taxa by habitat (morphologically similar creatures linked by brackets). (From Ralph Bulmer, "Why Is the Cassowary Not a Bird? A Problem of Zoological Taxonomy Among the Karam of the New Guinea Highlands," *Man*, (n.s.), 2, 1967. Reprinted by permission of the Royal Anthropological Institute of Great Britain and Ireland.)

agronomic activity." Included are such general categories as /lūta/ 'land, earth', and /bīlid/ 'mountainous slopeland,' as well as such specific categories as /quduŋo/ 'limonite,' and /qanul/ 'underground drainage conduits.' Among these are eight midrange categories that cover mutually exclusive surface types characteristic of all major vegetationally and agriculturally segregated lands. These are: /mapulun/, 'grassland'; /qinalāhan/, 'forest'; /mabilāu/, 'caneland'; /pinūgu/, 'woodlot'; /hābal/, 'swidden'; /lattaŋ/, 'house terrace'; /gīlid/, 'drained field in terrace'; and /payo/, 'pond field.' Various of these are leveled, tilled, left in woodland and/or otherwise managed maximally or intensively. Using the presence or absence of these four dimensions as semantic criteria, each of these eight surface types can be componentially defined as in Table 12.4.

According to Conklin (1967, p. 109), these categories also represent progressive and/or regressive stages in the development of agricultural

TABLE 12.4 IFUGAO LAND SURFACE CATEGORIES. [a]

G	1. mapalun	'grassland'	$\overline{L}$	$\overline{T}$	$\overline{W}$	$\overline{M}$
F	2. qinalāhan	'forest'	$\overline{L}$	$\overline{T}$	W	$\overline{M}$
C	3. mabilāu	'caneland'	$\overline{L}$	$\overline{T}$	$\overline{W}$	M
W	4. pinūgu	'woodlot'	$\overline{L}$	$\overline{T}$	W	M
S	5. hābal	'swidden'	$\overline{L}$	T		
H	6. lattaŋ	'house terrace'	L	$\overline{T}$		
D	7. gīlid	'drained field'	L	T		$\overline{M}$
P	8. payo	'pond field'	L	$\overline{T}$		M

L = leveled; T = tilled; W = wooded; M = managed maximally. $\overline{L}$, $\overline{T}$, $\overline{W}$, $\overline{M}$ = trait absent.

[a] *Data from Conklin, 1967, p. 108.*

lands; for example, F→S→W→P. Ideal patterns of land succession and use as perceived by the Ifugao can be matched to real patterns of land use. The categories can also be used by the investigator to record changes in the status of Ifugao lands through time. In the ecology of swidden agriculture, such an historical perspective can be particularly valuable as a measure of land-use intensity and productivity through time.

OTHER STUDIES IN ETHNOECOLOGY

Studies in ethnosystematics are essentially studies in the classificatory relationships among forms. As we have seen from the preceding examples, these relationships may be based on various criteria, including features of morphology, cultural use, habitat preferences, labor considerations, and so forth. The criteria need not be ecological in the sense of relating forms according to their interactions or according to their mutual influence on each other. Yet studies in "ethnoecology" using the strict ethnoscientific interpretation of the term should probe these areas of knowledge whether or not they are of classificatory significance in a given culture (Vayda and Rappaport, 1968, p. 491). Unfortunately, there are few studies in the literature that focus on these dimensions. However, a number of investigators report some data on native conceptions of ecological relationships in the context of other discussions. One brief example will be cited to illustrate this orientation.[15]

Warren Morrill (1967), in a report on the ethnoicthyology of the Cha-Cha, a people of French descent living on St. Thomas in the Virgin Islands, comments on several aspects of Cha-Cha knowledge of marine ecology. Morrill claims that the Cha-Cha know a great deal about the subject, and that they have a rather thorough working knowledge of the interrelationships of organisms on various types of reef formations and in

[15]See also Basso (1972); Bulmer (1965, 1968a); Forman (1967).

other marine environments. "The wide diversity of forms present on the reef is explained by ecological-evolutionary factors such as the long time required for reef formation, the complexity of shape (providing a large number of niches), and minor but significant differences in depth, clarity, currents and sunlight" (Morrill, 1967, p. 411). The Cha-Cha make these inferences based on *post hoc* observations of the distributions and behavior of the organisms rather than on some theory of their ecological relationships. For example, they observe that although lobsters normally live in sheltered areas, the female moves to areas where the water is turbulent at the time her eggs are ready to hatch. The Cha-Cha reason that this occurs because the baby lobsters must need more "air" in the water. They apparently do not recognize that the turbulent water scatters the young and reduces predation (Morrill, 1967, p. 411). Morrill also notes that the Cha-Cha understand the concept of a food chain with the exception of the role of microscopic organisms (to them invisible). According to their explanation,

nutrients ('chemicals') in the water are used by 'bugs' to synthesize food. The bugs are eaten by larger invertebrates or small fishes. These are taken by predators or 'just die' and become detritus eaten by invertebrates or return nutrients to the water (Morrill, 1967, p. 411)

Morrill adds several additional comments on Cha-Cha ecological views in the context of his presentation of taxonomies of reef and ocean fishes.

Studies in ethnoecology that attempt to demonstrate a correlation between human environmental behavior and native ecological and/or classificatory conceptions are also few, at least to date. Recall that Frake (1962a, p. 55)suggested that the method should be a first step in the development of predictive statements about the actual behavior of people with respect to their environments. He attempts to demonstrate such a correlation, at least in a preliminary way, as an adjunct to this initial methodological statement (Frake, 1962a). The case in point is an account of the ethnoecology of settlement patterns among the Eastern Subanun of Mindanao. Although recognizing that "this analysis should ideally rest upon a presentation of the pertinent Subanum ethnoscience relating to agriculture and vegetation types," Frake (1962a, p. 55) does not include these details in this general article in the interest of space.

The Eastern Subanum, a Philippine people, practice shifting farming in the tropical rain forests of Mindanao. In order to be successful, they "must in effect establish a controlled biotic community of sun-loving annuals and perennials in a climatic region whose natural climax community, the tropical rain forest, is radically different in almost every respect from the community agricultural man seeks to foster" (Frake, 1962a, p. 55). They achieve this end by periodically "putting the forest through its successional paces" rather than by replacing it with some wholly different biotic units, such as wet-rice paddies (Frake, 1962a, p. 55–56).

The Eastern Subanun settlement pattern is one of dispersed individual family households around swidden (farm fields) clusters. This contrasts with the dispersed swidden and nucleated village pattern maintained by many other Southeast Asian cultivators. Rules affecting this pattern for the Subanun derive in part from Subanun feelings about the relative amount of time and energy that should ideally be expended in such tasks as protecting swiddens from animal pests, building fieldhouses and traveling to fields, and in part from feelings about their proximity to neighbors. According to Frake (1962a, p. 57), they discuss actual and potential residence-site choices (location of houses) based primarily on these principles. They locate so as to maintain: (1) a minimum number of "wild vegetation boundaries" or areas 'to be watched'; (2) a minimum swidden-to-house distance for those with cultivating responsibilities; and (3) a maximum house-to-swidden distance from neighbors—so that family conversations and arguments cannot be overheard (Frake, 1962a, p. 57).

This practice actually results in some potential losses of farmland for the Subanun. According to Frake (1962a, p. 58), the Subanun emphasis on the immediate returns of swidden protection and accessibility, causes them to lose some control over the natural progression of forest revegetation. Maintaining narrow wild vegetational boundaries and dispersing houses tends to increase the size of cleared areas, thereby increasing the tendency for grassland rather than the needed forest to return. Other swidden farmers in other areas apparently weigh these and presumably other factors differently with different end results. Frake concludes:

The contrast among the remarkably different settlement patterns exhibited by Southeast Asian swidden farmers will become ecologically interpretable when one compares the factors which generate these patterns in each case rather than forcing ethnographic observations directly into a priori comparative categories . . . this incomplete account [for the Subanun] has revealed some of the advantages for the study of cultural ecology derivable from ethnographic description ordered according to the principles by which one's informants interpret their environment and make behavioral decisions. We are able to specify to what extent ecological factors determine settlement pattern, to point out significant general features of the

Subanun ecological adaptation, and to discover some meaningful dimensions for cross-cultural comparison. (1962a, p. 58)

In a recent paper on shifting farming among sharecroppers in Northeastern Brazil, Allen Johnson (1974) also attempts to demonstrate some concrete behavioral correlations to ecological data. Johnson's effort arises in part from criticisms that the ethnoecological approach should demonstrate in a more conclusive manner that the practical activities of a given group of people do bear some proper relationship to their ethnoscientific conceptions. In this paper, he attempts to demonstrate a statistical correlation between planting behavior by the Brazilian sharecroppers and their paradigm of "land types" for which rules as to which crops "like" which lands best can be explicitly stated (Johnson, 1974, p. 88). The land types terminologically distinguished include three swidden classifications (new, second-year, old), three ecological field zones (sandy hillside, river margin, river bottom) and two field types based on soil conditions (saline, and low, moist). These are categorized according to sharecropper views into a paradigm with two dimensions: hot (dry) *versus* cold (wet) and strong (fertile) *versus* weak (infertile). The strong–hot lands include the new and second-year swiddens. The strong–wet lands are the river margin and low, moist lands. The contrasting hot–weak lands are the old swiddens and the sandy hillside locations, while the cold–weak areas are confined to the river bottoms and saline lands.

Johnson (1974, p. 91) computed the amount and percentage of sharecropper fields according to this classification. He also recorded information from them as to the hot-cold-strong-weak preference of various commonly planted crops, including squash, manioc, beans, potatos, rice, and maize. He then counted the numbers of plants in the fields of six individual farmers and computed percentages of plants in each of four of the eight-field types. He found high agreement between the field classifications and actual planting practices and informants statements about the field requirements of plants (Johnson, 1974, p. 95). Johnson concludes that, although not perfect, this relatively simple paradigm and its associated data appear to account for the main trends in observed planting behavior.

CRITIQUE AND DISCUSSION

Ethnoscience and ethnoecology have been hailed by some as approaches that deserve serious attention by all ethnographers. Sturtevant (1964, p. 101) for example, says of ethnoscience, that it "shows promise as the New Ethnography required to advance the whole of cultural anthropology . . ." as it ". . . raises the standards of reliability, validity and exhaustiveness in ethnography." Others are less optimistic, noting that the results thus far have been more programmatic and methodological than concrete and explicative. Some question not only the techniques, but also the basic assumptions of both approaches.

Robbins Burling (1964), in one of the first statements critical of the methods of ethnoscience, raises two fundamental questions that remain as important today. The first is methodological and concerns the technique of componential analysis specifically. Burling (1964, p. 26) asks whether there cannot be many equally logical possibilities for analyzing the semantic dimensions of the same set of terms. If this is the case, then how is the investigator to "narrow the choice to the one or few that are 'psychologically real'?" (Burling, 1964, p. 26). Dell Hymes (1964) and Charles Frake (1964) answer Burling on this point, with Hymes in particular suggesting that although this may be the case in theory, it is probably not in practice. Hymes (1964, p. 117) notes that any analysis of terms according to semantic components should ideally be accomplished in the field so that it can be checked for correctness with informants and/or by independent observation of behavior. Any thorough analysis should also probe *why* category assignments are made as they are. If properly accomplished, this eliminates a number of the theoretical possibilities of an analysis and again places the task within the perspective of the culture being studied.

The other, and more fundamental of Burling's (1964) questions has to do with the implied relationship between linguistic elements (native terminologies) and cognitive processes. Are the two isomorphic? If not, what is their relationship? This is an old question that has plagued students of language and culture for more than a century. It is perhaps best known through the related formulations of the problem in the early 1930s by Edward Sapir and Benjamin Lee Whorf (Sapir-Whorf hypothesis), and it is still of concern today (Fishman, 1960). Burling (1964, p. 27) sees little in the methodology of ethnoscience that brings us closer to a solution of the problem. Others, including Paul Kay (1970), feel that the data of ethnoscience and general ethnographic semantics, when coupled with behavioral observations, do indeed contribute to our understanding of the complex relationship between language, thought and perception.[16] Berlin and Kay (1969) have also suggested that there may be

[16]The recognition by Berlin, Breedlove, and Raven (1974) that categories may be unmarked lexically is a case in point. See also Sapir (1912) for some early speculations on the role of language in environmental world views.

some universals operative in this relationship that deserve further study (see also Berlin, 1971).

Gerald Berreman (1966), in another critique, is of the opinion that most of the studies in ethnoscience to date have dealt with cultural trivia rather than ethnologically significant matters. He terms the analyses by Frake (1961) of Subanun disease terms, Conklin (1955) of Hanunoo color categories, Metzger and Williams (1966) of Tzeltal firewood types, and others, as examples of "anemic and emetic [rather than emic and etic] analyses" in anthropology. He also fears that the lure of this new "scientific" methodology may lead to many like contributions and also overshadow the gains to be made by other, more traditional approaches. To Berreman's "science of trivia" argument, A. Johnson (1974, p. 96) responds that the label "contains an indefensible element of ethnocentrism, for such domains [color, disease, firewood] are very important to the people being studied. If our task is to know another culture, then we may not regard as trivial those matters to which the culture ascribes importance." As to the lure of a false scientism, one can only remark that Frake (1962a), Conklin (1954) and others did not intend these approaches to totally replace existing methods. Frake explicitly states with reference to ethnoecology:

These methodological suggestions are not, of course, intended to replace the analysis of an ecosystem that Western biological science can provide. A scientific knowledge of the climate, soils, plants, and animals of a culture's environment is an essential foundation for ecological ethnography–but it does not, in itself, constitute ethnography. (1962a, p. 58)

Unfortunately, however, explications of the methods of ethnoscience and ethnoecology far outweigh the more thorough and complete studies that should be accomplished by combinations of these methods.

Marvin Harris (1968) in another critique is concerned with similar problems when he remarks on the seeming idealism of the approach. Harris fears that the implied stress on informants' verbal statements will lead to idealistic as opposed to realistic descriptions of cultural phenomena—ones of little direct importance to actual human behavior. Harris' concern with this specific problem has been countered by statements from Kay (1970, p. 21) to the effect that there are probably very few practicing ethnoscientists who take native generalizations at face value. The statements of informants always have to be compared to actual observed activities and other aspects of behavior. According to Kay, this is a requirement of all ethnographic methods, not just ethnoscience.

Vayda and Rappaport (1968) express a similar concern with the ethnoecological approach in particular. In addition to noting that most of the studies in this field to date have been what they term "ethnosystematic" rather than ethnoecological in the more restricted sense of the term, they also suggest that the approach has a "fundamental limitation for providing understanding of relations within ecological systems" (Vayda and Rappaport, 1968, p. 491). This limitation involves a difference between what might be called the "cognized environment" of a native people and their "operational environment" (Rappaport, 1963, p. 159). The cognized environment is the environment as understood by the people while the operational environment is the sum total of all environmental features whether fully comprehended or not. Vayda and Rappaport (1968, p. 491) suggest that the ethnoecological approach does not provide a mechanism for handling the operational environment should it prove—and it probably will—to be more complex than the cognized environment. They also cite several examples from the literature to indicate that there may be various cultural (e.g., sociological, religious) practices that have latent ecological functions in population dispersion, resource utilization, and so forth.

While the points made by Vayda and Rappaport (1968) are indeed important, Frake's (1962a) explicit statements as to the supplementary nature of ethnoecological approaches can also be recalled. Kay (1970, p. 22) also suggests that ethnoscience is often concerned with features for which informants do not have explanations and with factors of which they are unaware.

The common thread that seems to run through most of these criticisms is a legitimate concern for the relationship between culture—ideal and/or real—and behavior implied by these studies. As A. Johnson (1974, p. 97) has observed, the issue, while present by implication in most studies in ethnoscience and ethnoecology, has rarely been explicitly faced. The concern with demonstrating methods and systematics has been primary, perhaps because it was proposed as the logical first step by the methodologists. However, few seem to go on to the next step of independently observing behavior and then to the third of presenting rules for their correspondence (A. Johnson, 1974, p. 97). Until more studies of this kind are undertaken (as, for example, Johnson's), a full evaluation of the behavioral implications of the method cannot be made.[17] Also, until there are more thorough studies available in this area for ethnoecology, the full utility of the approach to the description of the

[17]M. Harris (1974b) would still be pessimistic. See also Keesing (1973) for a tongue-in-cheek comment on methods.

many and complex problems of human ecology cannot be assessed. However, as Frake (1962a, p. 53) has also proposed, ethnoecology need not be seen "simply as a methodological interest in its own right." Any ethnographer, ecologically oriented or not, who is interested in obtaining as complete a record of a given people as possible, cannot afford to neglect the thorough treatment of native conceptions. However, a consideration of the significance of these conceptions to their behavior—real and/or ideal—must also be thoroughly considered.

SUMMARY

This chapter has focused on the history, methodology, and criticisms of ethnoecology, a distinctive approach to human ecology that concerns itself with native conceptions of their environment. Its method, drawn principally from the method of ethnoscience, attempts to demonstrate the systematic relationships between native terminological systems for the environment and those conceptualizations. Data from several specific studies were presented to illustrate the procedures and some of the results of the method. Although a number of criticisms have been leveled at the approach, particularly for a seeming lack of concern with the behavioral implications of its method, the approach continues to hold promise if only as a means to more complete and thorough descriptions of the relationships of native peoples to their environments.

THIRTEEN HUMAN PALEOECOLOGY

The aim of this book is to show *how* ecological processes can be used to explain the origins and diversity of humans and their activities. By far the most complete information about human ecology comes from living groups, and this has been the focus of this book. Nevertheless, ecological approaches are being used to explain the fossil and archaeological records, and some examples have already been given. Because of gaps in our record of the past it is unlikely that this study will shed new light on ecological processes, with the exception of *rates* of change in ecological relationships. Studies of the past, however, can be used to *test* ecological hypotheses based on observations of living groups and can make an important contribution in this area. In this chapter the kinds of fossil and archaeological data that are relevant to ecological studies are discussed, along with the ways in which ecological approaches have been used to understand the past.

BASIC CONCEPTS OF PALEOECOLOGY

The *principle of uniformity* is the fundamental assumption of human paleoecology, as it is to all of the "paleo" sciences; that is, the present is a clue to the past. In geology the principle has long meant that the geological processes changing the face of the earth today are the same as those responsible for changes in the past. Its importance for identifying ancient environments is clear. The observation that the sediments deposited in lake bottoms are silt or clay helps to explain geological strata dominated by siltstone and claystone. Not only can the immediate environment of the strata be identified, but also the processes responsible for their deposition. The procedure is exactly the same in paleoecology. Plant and animal remains associated with geological strata are identified and their nonliving environment reconstructed from the sediments in the strata. The observation of similar plants and animals living together today under similar physical conditions is then used to explain their ecological relationships. In practice paleoecological data are used to *test hypotheses* about ecological relationships drawn from observations of today's world. Hypotheses are rejected or not according to how well they explain the data. The incompleteness of the fossil record, due to poor preservation most of all, and the immense complexity of known ecological relationships make the principle of uniformity essential to human paleoecology.

Stratigraphic associations are also essential to the study of extinct ecological relationships. That is, the remains of humans, plants, and animals, along with evidence of their nonliving environment, must be "sealed" together within a geological stratum. Each stratum is assumed to represent a distinct period of time and the remains contained within are, therefore, contemporaneous. Later disturbance of the stratum, perhaps mixing remains from several strata, would not, of course, allow such an assumption, and the possibility of disturbance must be carefully considered. Ecological relationships are recognized on different *scales*, as was pointed out earlier, so that stratigraphic associations must be viewed in several ways. A stratum at a single site may be appropriate for small scale or "microecology" studies, but strata from several sites must be correlated to understand ecological relationships on a larger scale. The problem of scale is particularly acute when mobile groups with different seasonal activities are being studied. In this case fossil data from several sites must be correlated to understand the ecology of a single population.

PALEOECOLOGICAL DATA

Ecological approaches to the study of the past require answers to two kinds of questions. The first question is what human populations existed and what plants, animals, climates, and so forth were associated with them. Unfortunately, many so-called ecological studies do not attempt to answer more than this question. The most important question is what were the environmental *relationships* into which the human populations entered? There has been, and still is, a tendency to confuse the identification of things in the environment with ecology. Three kinds of data in the fossil and archaeological records can be used to answer both questions: (1) human remains, (2) plant and animal remains, and (3) sediments.

Human Remains

The preserved material remains of humans and their activities are important clues to human ecology. On the one hand are the remains of dead individuals, including skeletal parts, fossilized or not, complete "mumified" humans, and earth "casts." On the other hand are tools, ornaments, graves, temples, and other material remains of human activities. *Skeletal material* provides the most direct clue to an ancient human population, but a number of problems hinder correct interpretation. The main difficulty is making sure that the sample of skeletons recovered from the archaeological site is representative of the original population. Archaeologists are seldom able to excavate complete sites so that the skeletal sample comes from only a portion of the site. There is no assurance that this sample can be extrapolated to the entire site; that is, excavation of 10 percent of the site does not necessarily mean that 10 percent of the skeletons have been recovered. Skeletons are often concentrated in a few areas of the site (cemeteries, ossuaries, etc.); if the excavation area of the archaeologist does not include these concentrations population estimates are bound to be too low. Assuming that the sample is representative of the population, we still cannot always be certain that all the dead are represented in the skeletons recovered. Cook (1972, p.5) mentions two main sources of error: adults who are killed at a distance while engaged in warfare or long-distance trading, and the very young whose bones disintegrate much more rapidly than adults. Finally, a third problem is encountered by the inability of skeletal material alone to give information on birth and migration rates, that is, what goes into the population (Cook 1972, pp. 6–7). Yet any estimation of actual size of a population at a given time must consider both. Consequently, other kinds of evidence must be used to estimate the fertility and migration components, perhaps by comparing size estimates obtained from settlement pattern studies and from cemeteries.

Archaeological *sites* can also be used to estimate population size *if* time can be controlled. Schwartz (1956) gives an example from the Grand Canyon in the American Southwest. An archaeological survey of parts of the Grand Canyon bottom located a group of sites. The occupation of each site was determined from the presence of painted trade wares, each known to have a definite, and short, time span. Site occupations were divided into 25-year "habitation units" on the basis of trade wares, and the number of sites occupied during each 25-year period was counted. Schwartz was able to show that population size in the Canyon was low during initial colonization between A.D. 600–700, grew rapidly from 700–900, stabilized from 900–1100, and finally declined rapidly from 1100–1200. This type of study is, of course, based upon *relative* population size and assumes that sites are of similar size.

The distribution of archaeological sites is another clue to ecological relationships of the past. Archaeological sites are varied, ranging from small temporary camps to large cities and give clues not only about where humans once lived, but also about their activities. Both kinds of clues are important to paleoecological studies. Once located, sites are identified by size and function and plotted on a distribution map. The arrangement of the sites vis-à-vis the landscape and each other, the *site* or *settlement pattern,* is then used to test hypotheses about ecological relationships, hypotheses drawn by analogy from living human groups. Chapters 8 and 9 suggest the kinds of relationships that have been or might be proposed to explain ancient population distributions. The most common hypotheses in paleoecology relate distributions to topography, use of microenvironments as determined by patterns of subsistence, disease zones, trade with other human groups, and warfare.

Artifacts are useful sources of paleoecological data. Cook (1972, pp. 10–11) gives the example of the plow in eleventh-century England. Historical information derived from the Domesday Book allowed one investigator to calculate that about 2.1 persons used each plow. Therefore, knowledge of the number of plows in an eleventh-century English town could be used to estimate the number of persons living in the town who were engaged in agricultural pursuits. Relationships between human populations can sometimes be inferred from the presence of *trade goods.* Trading is recognized in the remains of ancient settlements either by the presence of raw materials that are not now locally available or by manufactured objects that are stylistically, technologically, or otherwise unique (Wright, 1974). Once identified, trade goods must be traced as closely as possible to their origin. Cross-cultural comparison, that is, comparing the style or technology of the object to that of objects recovered from other settlements, is one method. Thus trade between the Mexican city of Teotihuacan and the Classic Maya was recognized from the occurrence of typical Teotihuacan pottery vessels and architectural styles in Maya sites. Other methods make use of physical and chemical analysis. In recent years one of the most popular methods is to identify the *trace elements* in the traded object, that is, elements making up less than 0.1 percent of the material, and to construct a characteristic "fingerprint." The trace element fingerprint is then compared to those from various sources of the material suspected to be the origin. Obsidian, a glassy volcanic stone, was shown by this method to have been traded widely throughout the Mediterranean from only two sources in Turkey (Wright, 1974).

Population size can be estimated from the calculation of the amount of *space* covered by rooms, houses, and sites (Cook, 1972, pp. 12–23). This approach is based upon the assumption that general rules can be found that relate the amount of dwelling space to the number of occupants. Ethnographic analogy is used to formulate the rules based on three kinds of information: (1) the type of social unit commonly inhabiting a room, house, or site, for example, nuclear family, extended family, lineage; (2) the number of persons commonly found in the social unit; and (3) the amount of habitation space commonly used by the social unit. Most studies have stressed the nuclear family. Cook (1972, pp. 13–15) suggests that studies of single, one-room dwellings in the United States show that nuclear families occupied 120 to 350 square feet of space. Each nuclear family was found to include an average of 5.0 persons but almost never more than 6.0 persons. The general rule, then, is that a house with a floor space of less than 350 square feet was probably not occupied by more than 6 persons. [However, the demographer William Petersen (1975) points out that the number of people per nuclear family, per room, and floor space per person or family is subject to great variation and averages may be meaningless.] Similar procedures have been used to find the relationship between larger social groups and habitation space. The total size of an ancient population can then be calculated if (1) all the rooms and houses occupied by the population for a given time period have been located or their numbers accurately estimated; and (2) all of the rooms and houses in existence during a given time period were actually occupied by members of the population or, if not, the proportion actually occupied is known.

Plant and Animal Remains

The remains of plants and animals give direct clues to ancient environments and, when associated with human coprolites, stomach contents, or features of human origin (e.g., hearths or storage pits), can be used to identify the ecological relationships into which humans enter. Paleoecological data of this kind are most often associated with stream, lake, swamp and spring beds, beach and estuarine beds, loess and volcanic ash, dry caves, and archaeological sites (Butzer, 1971, p. 254). With the exception of "megafossils," such as the bones of large mammals or construction timbers, plant and animal remains are apt to be overlooked in archaeological deposits so that specialized recovery techniques are used. Passing the sediments containing the remains through fine mesh screens is perhaps the most common method. Often a series of screens is used, starting with a large mesh (e.g., around ½ inch) and passing the sediments through successively smaller screens, down to 1/16 inch or even smaller. This is a time-consuming method, and sampling schemes are used to get a representative sample, rather than screen all of the deposit. Alternatively, a medium-size screen is used (e.g., around 1/4 inch) to screen all of the deposit; the reliability of the method is checked by passing samples of the sediment through smaller mesh screens. *Water* and *chemical flotation* are other methods that have been developed to improve the recovery of macrofossils (Streuver, 1968a). Both depend upon differences in specific gravity to separate one kind of material from another. Water flotation is used to separate out heavy inorganic materials from light plant remains and animal bones. The sediment is dumped, a little bit at a time, into a tub filled with water; the heavy material sinks rapidly to the bottom while the lighter material sinks much more slowly, allowing someone to skim it off and remove it to another container. Chemical flotation is used to separate bone from plants, again on the basis of differences in the specific gravity of the two. A solution of zinc chloride is put into a small container and mixed plant and animal remains added. The heavier bone sinks to the bottom while the light plant remains afloat at the top, allowing removal to a separate container.

POLLEN ANALYSIS. The way in which plant and animal remains are used to reconstruct ancient environments and ecological relationships is well illustrated by plant *pollen*. Pollen grains are produced in large quantities by plants as part of the reproductive process, are resistant to decay because of a waxy outer coating, and are about 25 microns in diameter. The grains have a complex and variable morphology that allows identification, by genus or sometimes species, of the plant that produced them. For these reasons the study of pollen, *palynology*, has become one of the more important ways of recognizing ancient plant environments. The recovery of fossil pollen is done by taking samples of sediments and preparing the sample in the laboratory so that the pollen grains can be removed and identified. Karl Butzer suggests the following procedure:

Samples should be removed from stratigraphically meaningful horizons or sections at vertical intervals of 5, 10, 20, or 30 centimeters as the case warrants. Some 20 cubic centimeters of material will suffice in the case of all but sands or deeply weathered sediments. The sample should be removed with great care so as to avoid contamination from the atmosphere or the removing tool, preferably after cleaning and discarding the surface layer. Short test tubes with a cap are most convenient for storage. In the case of general pollen studies applied to nonexposed strata, such as lake or bog sediments, core borings are made using various hand-operated devices. (1971, p. 244)

Preservation of the pollen grains affects their recovery. Not only is the pollen of some plants more easily preserved but some sediments are more conducive than others for preservation. Thus organic and semiorganic sediments of lacustrine origin are excellent for preservation; clays and dense silts (if not deeply weathered) are favorable; and sands are poor (Butzer, 1971, p. 244).

Fossil pollen allows directly only identification of ancient plants, not an understanding of ecological interrelationships. Palynologists use the principle of uniformity to infer the latter (Davis, 1969; Dimbleby, 1967). The pollen "rain" from modern plant communities is collected from the surface and used to make a pollen "fingerprint" of the community. The fingerprint is based on percentages and ratios of one kind of pollen to another. Pollen percentages are not the same as plant percentages in the community because plants produce different amounts of pollen, pollen with different dispersal characteristics, and pollen that preserves differently. However, each community does tend to produce distinctive pollen percentages so that a pollen "profile" based upon them can be used as an identifying fingerprint or signature. Once the fingerprint has been recognized, palynologists can then turn to the fossil record and compare fingerprints. If the ancient environment contains plant communities similar to a modern one, ancient ecological relationships can be inferred from the modern community.

CLIMATIC INDICATORS. Plant and animal remains are also used as clues to ancient climates and other aspects of the nonliving environment. The faunal assemblages from the Pleistocene have long been used to support the hypothesis that the Pleistocene was a cold period and was marked by several warm/cold cycles. The megafauna of Upper Pleistocene Europe is a good example. Most of the fauna in the European assemblage can be observed today (only 3 genera have become extinct) and many tundra species are present, including reindeer, musk, ox, snowshoe and artic fox (Butzer, 1971, pp. 258–59). To these can be added the now extinct woolly mammoth and woolly rhino, both of which are known to be cold-adapted from nearly complete "mummified" remains. There can be little question that midlatitude Europe was much colder during the Pleistocene than it is now. However, Upper Pleistocene climatic conditions cannot be simply equated with the modern arctic tundra because the assemblage also includes a cool, midlatitude steppe fauna, such as the saiga antelope, wild steppe horses, the steppe fox, polecat, and marmot, the hamster, and a gerbil (Butzer, 1971, pp. 266–67). The most important climatic indicators are one-celled organisms called

Foraminifera that live near the ocean surface and are very sensitive to temperature changes. Since the temperature of the ocean surface water reflects regional climates, fossil Foraminifera are good indicators of climatic changes. Pleistocene warm/cold cycles can then be interpreted from warm and cold Foraminifera assemblages recovered from the deep sea sediment cores (Butzer, 1971, pp. 266–67).

Rainfall is another climatic variable that has been estimated from fossil plants and animals. One of the classic examples is the analysis of the Paleolithic-Mesolithic faunal remains at Mount Carmel (Israel) by D. M. A. Bate in 1937 (Garrod and Bate, 1937, pp. 139–53). The relative frequency of fallow deer remains and gazelle remains in the Mount Carmel strata were used by Bate to infer changes in rainfall. Fallow deer are woodland animals while gazelles inhabit a dryer, even desert, country. An increase in deer remains would mean an increase in woodlands and presumably an increase in rainfall and vice versa. Of course changes in the proportion of deer to gazelle remains could also be interpreted as reflecting a change in the food preferred by the stone age hunters, and Bate's interpretation has been so criticized. Nevertheless, the general method is still valuable and selection of the best interpretation must depend upon other data.

As a final example, stream and lake characteristics can be interpreted from the remains of freshwater mollusca. The earliest use of mollusks for that purpose of which I am familiar is by J. P. E. Morrison (1942). Morrison examines freshwater mollusks from several kitchen midden mounds in northwest Alabama and finds that most of the species are adapted to shallow shoals although a few are from deeper water. Accordingly, he argues that conditions of the Tennessee River running near the mounds were the same at the time of human occupation as they are now. In general freshwater mollusks can be grouped into the the following kinds of indicators of water conditions (Sparks, 1963, p. 322):

1. Slum species—those that can survive small, shallow bodies of water with poor aeration, periodic drying, and large temperature changes.

2. Catholic species—those that are found in a wide variety of freshwater habitats.

3. Ditch species—those that are adapted to slow moving streams.

4. Moving water species—those that are adapted to large, moving bodies of water.

CLUES TO HUMAN ECOLOGY. Only under some circumstances can plant and animal remains be used to infer the ecological relationships into which humans enter. Containment within an archaeological "feature"—a distinct physical structure such as a hearth, pit, house floor, or pyramid—is the strongest evidence. For example, the presence of charred seeds in the bottom of a pottery vessel or broken, burned animal bones in the ash of a hearth suggest a predator-prey relationship, that is, the relationship that exists between a human population and its food supply. Macrofossils recovered from *coprolites,* the preserved fecal remains of animals, points to a similar relationship (Bryant and Williams-Dean, 1975; Callen, 1969). Coprolites are most frequently preserved in dry caves or rock shelters and for that reason coprolite studies are becoming common in arid areas such as the desert west of the United States. Human coprolites give direct evidence of the plants and animals that were eaten and thus can be very useful in pinpointing ecological relationships. Since coprolites are seldom abundant, they may not reflect the typical diet of the human population as a whole. Once again the problem of representativeness haunts the paleoecologist. In order to successfully use data from coprolites for studying ecological relationships, it is necessary to compare with as many other sources of ecological data as possible.

Finally, a few rare data sources should be mentioned. Plant and animal remains have been found in the stomach and intestines of "mummies," including the so-called "bog people" of Europe and the Egyptian mummies. For example, study of the internal viscera of the famous Tollund man of Denmark showed that although the contents of the stomach consisted of vegetable remains of a gruel prepared from barley, linseed, "gold-of-pleasure" *(camelina sativa)* and knotweed, with many different sorts of weeds that grow on ploughed land, it could not have contained any meat at the time of death, since recognizable traces of bone, sinew, or muscular tissue would certainly have remained (Glob, 1971, p. 12). Other rare sources include imprints of plant parts on pottery containers and clay or mud bricks, floor material, and so forth, carbonized grain from ancient drying kilns, hearths, or cooking containers, and the "mummified" grain from storage bins between the Egyptian pyramids (Helbaek, 1969).

PROBLEMS OF INTERPRETATION. Interpreting plant and animal remains in archaeological deposits is plagued with problems that are both familiar and unknown to paleoecologists working with nonhuman remains. On the one hand, archaeological deposits are subject to the same physical processes that affect any kind of material object, whether that object be a particle of sand, a boulder, a ring of stones around what was once a house, or the bony remains of a butchered animal. The processes, which may include water flow, wind and water erosion, and fracturing due to temperature changes, among other things, not only changes the shape or composition of an object, but also may cause movement and, consequently, changes in the way that objects are aligned in the deposit. Geologists interested in the deposition of sedimentary particles have used observations of modern physical processes to develop models that realistically explain how geological deposits were formed (Society of Economic and Mineralogic Petrologists, 1972). The recent development of *taphonomy,* the study of the laws of burial, reflects the application of this approach to paleontological and archaeological deposits. The goal of taphonomy is to understand the processes taking place between the death of an organism and the deposition of its remains in a geological deposit (Gifford, 1975). Of particular interest are those processes that result in the *movement* of remains from the place of death to the final resting place. Movement of some kinds of remains and not others can greatly affect the interpretation of ancient ecological relationships. Other processes affect the chances that remains will be preserved and also have a great impact on the correctness of an interpretation.

On the other hand, plant and animal remains in archaeological deposits are in large part selected through a "cultural filter." The remains are in the deposit because of human activities and do not necessarily represent the ancient environment of the human population. Long-distance trade may be responsible for some of the remains, and selective hunting and gathering of plants and animals in the vicinity may give remains that create an erroneous impression of the environment. Furthermore, a change in diet can take place, causing a change in the deposit's plant and animal remains, without a corresponding shift in the ancient environment. The implications of these data for paleoecology must be carefully considered.

Sediments

An important source of information about ancient physical environments, e.g., climate, is the study of *sediments* in which fossil and archaeological remains are found. Sediments are rock particles that have been detached by weathering and transported to a final resting place by gravity, running water, wind, or glaciation. The processes of weathering and transportation have predictable effects upon the particles, making sediments clues to past environments. Sedimentary composition, texture, and structure are the principle clues and should be carefully studied by specialists (Ager, 1963, pp. 123–36). The size and shape of sedimentary particles have been particularly useful as climatic indicators. Thus Francois Bordes's (1972) work on Middle Paleolithic deposits in French caves has suggested that rock fragments detached from the roof and walls of caves during cold periods have sharp, angular edges while those falling during warm periods have weathered, rounded edges. An adequate treatment of sediments is not possible here; the interested reader should consult Karl Butzer's (1971) *Environment and Archaeology*, 2nd ed., or the recently published *Archaeological Sediments* by M. L. Shackley (1975) for an introduction to the use of sediments in human paleoecology.

ECOLOGICAL APPROACHES TO THE PAST

The preceding section has outlined the kinds of data in the fossil and archaeological records that can be used for ecological studies. Paleoecological data have been used to test ecological models explaining the past. The explanation of human diversity and the origin of new human organizations has been the fundamental goal. However, human paleoecological data have also been used to test models of long-term changes in large-scale ecological systems.

Diversity

Variability in the fossil and archaeological records is no less perplexing than it is in living populations and, not surprisingly, ecological explanations have been used to alleviate the problem. While no two fossil hominids are exactly alike, those belonging to a single biological taxon, for example, a species, are assumed to be more like each other than members of another taxon. Grouping together fossil individuals into a limited number of taxa, then, is the procedure used to explain diversity in the fossil record. Unfortunately, there is little agreement about how much and what kind of variation can be used to recognize a given taxon. The problem is just as real with living taxa but here there is recourse to observed reproductive behavior and other characteristics that are not fossilized. Identifying a few traits with "taxonomic relevance," that is, traits that are associated *only* with individuals of a given taxon and can, therefore, be used to separate one taxon from another, is essential to the paleoanthropologist. Some characteristics of this sort can best be explained as "novelties" established in a population by chance, for example, a crescent-shaped depression at the base of the skull in some Middle

Mississippian populations in the southeastern United States. Others, however, are explained by ecological hypotheses, particularly environmental adaptation. Thus the taxonomic separation of the "classic" Neanderthals in Europe from the "progressive" Neanderthals in Middle East is based upon a complex of biological adaptations to cold weather in the former. Classification of the australopithecines is another case in point. These early hominids are commonly separated into two taxa, a *gracile* and a *robust* group, on the basis of distinctive dentition, jaw size, and musculature. The differences, according to one hypothesis, can best be explained as dietary adaptations, the gracile group having an omnivorous diet and living in grasslands while the robust group lived in the woodland fringes and was herbivorous (see Swedlund, 1974, for a good discussion).

Ecological models are being used more and more to explain variability in the archaeological record. The American archaeologists Sally Binford and Lewis Binford (1966) give a classic example from the Middle Paleolithic of Europe and the Near East. The Mousterian assemblages of this period are marked by a distinctive set of stone tools but vary in the proportions of each tool type making up an assemblage. The French prehistorian Francois Bordes (1972) argues that the variation can best be explained as *stylistic,* representing several ethnic groups with somewhat different cultural traditions. On the other hand, the Binfords propose that the variation in Mousterian assemblages is better explained as *adaptive* responses made by a single ethnic group to a heterogeneous environment. That is, distinctive assemblages represent tool kits associated with activities used to extract and prepare different resources, resources that vary from place to place and by season. The Binfords used a method of statistical analysis, factor analysis, to identify five distinct tool kits from the Mousterian assemblages. The tool kits suggested the range of activities expected by the "adaptive" model and gives support to it. At the present time the two competing explanations are at a standoff. Bordes has pointed out data suggesting that some of the Mousterian sites were occupied year-round, clearly supporting his model, but Lewis Binford (1973) feels that the evidence is inconclusive because of the methodology employed. Until this question can be resolved, and other data brought to bear, the standoff will continue and will give testimony to the difficulties of explaining variability in the archaeological record.

Origins

Ecological models have also been used to explain the origin of new biological and sociocultural organizations. The first use is exemplified by Clifford Jolly's (1970) model of the emergence of a distinctive hominid dentition pattern. Jolly argues that hominid dentition differs from that of pongids (apes) in the same way that the dentition of the Ethiopian highland gelada, *Theropithecus gelada,* differs from its close relative *papio,* the savannah baboon. The differences are functionally related dental characteristics that he calls the "T-complex," including small canines, vertical incisors, a distinctive mandible shape, and mesial drift (teeth crowding toward the front of the jaw with age). "T-complex" dentition, according to this model, is an adaptation to a diet of small, hard seeds, tough grasses, and other plants requiring constant and forceful grinding. Jolly sees the hominid dental pattern originating in an ancestor that was adapted to a similar diet (he calls it "granivorous"), in contrast to an alternative, and more traditional, model that explains reduced canine size, and so forth, as the consequence of learning to use tools. Both *Gigantopithecus,* a fossil hominoid from Asia, and *Ramapithecus,* the earliest fossil hominid, have "T-complex" dentition and support Jolly's model. (See, however, Wolpoff, 1971, for a contrary view.)

The second use is exemplified by a proliferating group of models proposed by archaeologists and anthropologists to explain the origins of domestication, the state, civilization, and other novel sociocultural patterns. Many of these have already been discussed. The population pressure model of the origins of domestication in the Near East, proposed by Lewis Binford (1968) and developed by Kent Flannery (1969), is typical. Another is William Rathje's (1971) explanation for the origin of lowland Classic Maya civilization.

Testing Models of Change in Ecological Systems

A final use of paleoecological data is in testing ecological models about long-term changes in ecological systems. Betty Meggers (1975) gives an excellent example from the tropical lowlands of South America. Ecologists have long been puzzled by the contradiction between the tremendous diversity of plants and animals in tropical South American and the lack of natural barriers usually associated with the origin of species. In recent years, however, a model of diversification based upon cycles of climatic change has been proposed. The model assumes that several times in the past the intensity of rainfall was reduced, causing the replacement of tropical forest by savannahs in all but a few "refuge" areas. Since most plant and animal species were forest-adapted, the savannahs were effective barriers to movement among the forest refuges and established good conditions for speciation. Several cycles of forest fragmentation and subsequent coalescence have been proposed, two of which take place within the past 12,000 years and after the arrival of man.

Meggers shows that the distribution of linguistic and ethnographic groups in the region could be explained by the model, but the most direct evidence comes from human remains. Archaeological sites throughout the South American tropics are highly variable, suggesting human groups that have been isolated from each other and consistent with the idea of scattered forest enclaves separated by savannah barriers. Meggers points out that the savannahs are notoriously poor places for humans to find food and would present real obstacles to movement, particularly to forest-adapted groups. In the center of the region archaeological sites are marked by remains of several distinct groups separated by gaps in the record. Thus the Ucayali region in eastern Peru was successively occupied by 12 such groups and Marajo Island at the mouth of the Amazon by 5. The remains of many of the groups are widespread, as evidenced by several ceramic styles distributed at various time periods from eastern Peru to the mouth of the Amazon. All in all these data suggest the rapid movement of culturally distinct human groups into and throughout the central region, consistent with the hypotheses of reforestation during a period of increased rainfall in the climatic cycle. That is, with increased rainfall savannahs would be replaced by tropical forest, eliminating the barriers to movement and allowing once-isolated human groups to expand. Topographically, the South American tropics are shaped like a soup plate and the postulated forest refuges are in the higher, peripheral regions. Reforestation would funnel expanding populations into the lower central region. The diversification taking place in the refuges, therefore, would be reflected in the center of the zone and is indeed found in the archaeological record.

SUMMARY

The fossil and archaeological record is a source of data that can be used to test ecological models explaining human diversity in the past and the origins of new organizations. Human "paleoecology" assumes the principle of uniformity and adequate control of the time dimension. Given the validity of these assumptions, ecological studies can be based upon human remains, plant and animal remains, and sediments. However, it is important to recognize when these data only give clues to the existence of a human population and its environs and when they give clues to ecological relationships. Only in the latter case can studies of human paleoecology be carried to fruition. In most cases ecological explanations of the past depend upon models of ecological processes taking place at the present. Paleoecological data are simply not adequate to do more than provide tests of these models. However, the fossil and archaeological record can give information about long-term ecological processes not observable at the present, particularly the rates at which ecological relationships change over time.

REFERENCES

Adams, R. M., *Land Behind Bagdad: A History of Settlement on the Diyala Plain*, University of Chicago Press, Chicago, 1965.

———. *The Evolution of Urban Society*, Aldine, Chicago, 1966.

Ager, D. V., *Principles of Paleoecology*, McGraw-Hill, New York, 1963.

Alexander, R. D., "The Evolution of Social Behavior," *Annual Review of Ecology and Systematics*, 5: 325–383 (1974).

Allan, W., *The African Husbandman*, Oliver and Boyd, Edinburgh, Scotland, 1967.

Alland, A., *Adaptation in Cultural Evolution: An Approach to Medical Anthropology*, Columbia University Press, New York, 1970.

Anderson, E. N., Jr., "The Life and Culture of Ecotopia," In D. Hymes, ed., *Reinventing Anthropology*, Random House, New York, 1972, pp. 264–283.

Anderson, J. N., "Ecological Anthropology and Anthropological Ecology." In J. J. Honigman, ed. *Handbook of Social and Cultural Anthropology*, Rand McNally, Chicago, 1973, pp. 179–239.

Arensberg, C. M., and S. T. Kimball, *Family and Community in Ireland*, 2nd ed., Harvard University Press, Cambridge, Mass., 1968.

Arnheim, N. and C. E. Taylor, "Non-Darwinian Evolution: Consequences for Neutral Allelic Variation," *Nature*, 223: 900–903 (1969).

Aschmann, H., *The Central Desert of Baja California: Demography and Ecology*, Ibero-Americana 42, 1959.

Barrau, J., "L'Humide et le Sec, An Essay on Ethnobiological Adaptation to Contrastive Environments in the Indo-Pacific Area," *Journal of the Polynesian Society*, 74: 329–346 (1965).

Barth, F., "Ecologic Relationships of Ethnic Groups in Swat, North Pakistan," *American Anthropologist*, 58: 1079–1089 (1956).

———. *Nomads of South Persia*, Little, Brown, Boston, 1961.

Basso, K. H., "Ice and Travel Among the Fort Norman Slave: Folk Taxonomies and Cultural Rules," *Language in Society*, 1: 31–49 (1972).

Bates, D., "The Role of the State in Peasant-Nomad Mutualism," *Anthropological Quarterly*, 44: 109–31 (1971).

Bates, M., "Human Ecology." In A. L. Kroeber, ed., *Anthropology Today*, University of Chicago Press, Chicago, 1953, pp. 700–13.

Bateson, G., "The Role of Somatic Change in Evolution," *Evolution*, 17: 529–39 (1963).

Benedict, B., "Social Regulation of Fertility." In G. A. Harrison and A. J. Boyce, eds. *The Structure of Human Populations*, Oxford University Press, New York, 1972, pp. 73–89.

Bennett, J., "Anthropological Contributions to the Cultural Ecology and Management of Water Resources." In L. D. James, ed., *Man and Water, the Social Sciences in Management of Water Resources*, University of Kentucky Press, Lexington, 1974, pp. 38–41.

Berlin, B., "Speculations on the Growth of Ethnobotanical Nomenclature," *Language in Society*, 1: 51–86 (1971).

———, D. E. Breedlove, and P. H. Raven, *Principles of Tzeltal Plant Classification: An Introduction to the Botanical Ethnography of a Mayan-Speaking People of Highland Chiapas*, Academic Press, New York, 1974.

———, and P. Kay, *Basic Color Terms: Their Universality and Evolution*, University of California Press, Berkeley, 1969.

Berg, A., *The Nutrition Factor*, Brookings Institution, Washington, D.C., 1973.

Berreman, G. D., "Anemic and Emetic Analyses in Social Anthropology," *American Anthropologist*, 68(2, pt. 1): 346–54 (1966).

Binford, L. R., "Post-Pleistocene Adaptations." In S. R. Binford and L. R. Binford, eds., *New Perspectives in Archaeology*, Aldine, Chicago, 1968, pp. 313–41.

———, "Interassemblage Variability—The Mousterian and the 'Functional' Argument." In C. Renfrew, ed., *The Explanation of Culture Change: Models in Prehistory*, University of Pittsburgh Press, Pittsburgh, Pa., 1973, pp. 227–54.

———, and S. R. Binford, "A Preliminary Analysis of Functional Variability in the Mousterian of Levallois Facies." in J. D. Clark and F. C. Howell, eds., *Recent*

Studies in Palaeoanthropology, Special Publication of the American Anthropologist, 68(2, pt. 2): 238–95, 1966.

Birdsell, J., "Some Environmental and Cultural Factors Influencing the Structuring of Australian Aboriginal Populations," *American Naturalist,* 87: 171–207 (1953).

———, *Human Evolution,* 2nd ed., Rand McNally, Chicago, 1975.

Bordes, F., *A Tale of Two Caves,* Harper & Row, New York, 1972.

Borgstrom, G., "Ecological Aspects of Protein Feeding: The Case of Peru." In M. T. Farvar and J. P. Milton, eds., *The Careless Technology: Ecology and International Development,* Natural History Press, Garden City, N. Y., 1972, pp. 753–74.

Boserup, E., *The Conditions of Agricultural Growth,* Aldine, Chicago, 1965.

Boughey, A. S., *Man and the Environment,* 2nd Ed., Macmillan, New York, 1975.

Braidwood, R., and C. Reed, "The Achievement and Early Consequences of Food Production," *Cold Spring Harbor Symposia on Quantitative Biology,* 22: 19–31 (1957).

Bright, J. O., and W. Bright, "Semantic Structures in Northwestern California and the Sapir-Whorf Hypothesis." In E. A. Hammel, ed., *Formal Semantic Analysis, Special Publication of the American Anthropologist,* 67(5, pt. 2): 249–58 (1965).

Bronson, B., "Farm Labor and the Evolution of Food Production." In B. Spooner, ed., *Population Growth: Anthropological Implications,* MIT Press, Cambridge, Mass., 1972, pp. 190–218.

Bryant, V. M., and G. Williams-Dean, "The Coprolites of Man," *Scientific American,* 232: 100–109 (1975).

Buckley, W., *Sociology and Modern Systems Theory,* Prentice-Hall, Englewood Cliffs, N.J., 1967.

Bulmer, R., "A Primitive Ornithology," *Australian Museum Magazine,* 12: 224–29 (1957).

———, "Beliefs Concerning the Propagation of New Varieties of Sweet Potato in Two New Guinea Highlands Societies," *Journal of the Polynesian Society,* 74: 237–39 (1965).

———, "Why Is the Cassowary Not a Bird? A Problem of Zoological Taxonomy Among the Karam of the New Guinea Highlands," *Man* (n.s.), 2: 5–25 (1967).

———, "The Strategies of Hunting in New Guinea," *Oceania,* 38: 302–18 (1968a).

———, "Worms That Croak and Other Mysteries of Karam Natural History," *Mankind,* 6: 621–39 (1968b).

———, "Which Came First, the Chicken or the Egghead?" In J. Pouillon and P. Maranda, eds., *Echanges et Communications, Melanges Offerts a Claude Levi Strauss a l'occasion de son 60 eme Anniversaire,* Mouton, the Hague, 1970, pp. 1069–1091.

———, and M. Tyler, "Karam Classification of Frogs," *Journal of the Polynesian Society,* 77: 333–85 (1968).

Burling, R., "Cognition and Componential Analysis: God's Truth or Hocus-Pocus?" *American Anthropologist,* 66: 20–28 (1964).

Burrows, E., "Breed and Border in Polynesia," *American Anthropologist,* 41: 1–21 (1939).

Butzer, K., *Environment and Archaeology,* 2nd ed., Aldine, Chicago, 1971.

Cain, S. S., "Biotope and Habitat." In F. F. Darling and J. P. Milton, eds., *Future Environments of North America,* Natural History Press, Garden City, N.Y., 1966, pp. 38–54.

Callen, E. O., "Diet as Revealed by Coprolites." In D. Brothwell and E. S. Higgs, eds., *Science in Archaeology,* rev. ed., Thames and Hudson, London, England, 1969, pp. 235–43.

Carneiro, R., "Slash and Burn Agriculture: A Closer Look at Its Implications for Settlement Patterns." In A. F. C. Wallace, ed., *Men and Cultures: Selected Papers of the Fifth International Congress of Anthropological and Ethnological Sciences,* University of Pennsylvania Press, Philadelphia, 1960, pp. 229–34.

———, "On the Relationship Between Size of Population and Complexity of Social Organization," *Southwestern Journal of Anthropology,* 23: 234–43 (1967).

———, "A Theory of the Origin of the State," *Science,* 169: 733–38 (1970).

Casteel, R., "A Re-examination of Environmental Factors and Hunter-Gatherer Tribal Areas," Unpublished Paper Given at the 1974 Great Basin Anthropological Conference, Carson City, Nev., 1974.

Castetter, E. F. and W. H. Bell, *Pima and Papago Indian Agriculture,* University of New Mexico Press, Albuquerque, 1942.

Chagnon, N., *Yanomamo: The Fierce People,* Holt, Rinehart, and Winston, New York, 1968.

———, *Studying the Yanomamo,* Holt, Rinehart and Winston, New York, 1974.

Childe, V. G., *Social Evolution,* C. A. Watts, London, England, 1951.

Chomsky, N., *Aspects of the Theory of Syntax,* MIT Press, Cambridge, Mass., 1965.

Clarke, W. C., "From Extensive to Intensive Shifting Cultivation," *Ethnology,* 5: 347–59 (1966).

Coe, M. D., and K. Flannery, "Microenvironments and Mesoamerican Prehistory," *Science,* 143: 650–54 (1964).

Cohen, J., "Alternative Derivations of a Species—Abundance Relation." *American Naturalist,* 102: pp. 165–172 (1968).

Colby, B. N., "Ethnographic Semantics: A Preliminary Survey," *Current Anthropology,* 7: 3–17 (1966).

Colinvaux, P., *Introduction to Ecology,* John Wiley & Sons, 1973.

Colwell, R. H., and D. J. Futuyma, "On the Measurement of Niche Breadth and Overlap," *Ecology,* 52: 567–76 (1971).

Commoner, Barry, *The Closing Circle,* A. Knopf, New York, 1971.

Conklin, H. C., "An Ethnoecological Approach to Shifting Agriculture," *New York Academy of Sciences, Transactions,* 17(2): 133–42 (1954).

———, "Hanunoo Color Categories," *Southwestern Journal of Anthropology,* 11: 339–44 (1955).

———, *Hanunoo Agriculture: A Report on an Integral System of Shifting Cultivation in the Philippines,* Food and Agricultural Organization, United Nations, Rome, Italy, 1957.

———, "Lexicographical Treatment of Folk Taxonomies," in F. W. Hoseholder and S. Saporta, eds., *Problems in Lexicography,* Publications of the Indiana University Research Center in Anthropology, Folklore, and Linguistics, No. 21, 1962, pp. 119–41.

———, "Some Aspects of Ethnographic Research in Ifugao," *New York Academy of Sciences, Transactions,* 30: 99–121 (1967).

———, *Folk Classification: A Topically Arranged Bibliography of Contemporary and Background References Through 1971*, Department of Anthropology, Yale University, New Haven, 1972.

Cook, S. F., "Human Sacrifice and Warfare as Factors in the Demography of Pre-Colonial Mexico," *Human Biology*, 18: 81–102 (1946).

———, *Prehistoric Demography*, McCaleb Module 16, Addison-Wesley, Reading, Mass., 1972.

Cottrell, F., *Energy and Society*, McGraw-Hill, New York, 1955.

Coursey, D. G., and C. K. Coursey, "The New Yam Festival of West Africa," *Anthropos*, 66: 44–48 (1971).

Cowgill, G. L., "On Causes and Consequences of Ancient and Modern Population Changes," *American Anthropologist*, 77, pp. 505–525 (1975).

Cravioto, J., E. R. Delicardie, and H. G. Birch, "Nutrition, Growth and Neurointegrative Development: An Experimental and Ecological Study," *Pediatrics*, 38: 319–72 (1966).

Damas, D., "Environment, History, and Central Eskimo Society." In D. Damas, ed., *Contributions to Anthropology: Ecological Essays*, National Museum of Canada, Bulletin No. 230, Ottawa, 1969, pp. 40–64.

Darwin, C., *On the Origin of the Species by Means of Natural Selection or the Preservation of Favored Races in the Struggle for Life*, Murray, London, England, 1859.

Davis, M. B., "Palynology and Environmental History During the Quaternary Period," *American Scientist*, 57: 317–32 (1969).

Deevey, E. S., "The Human Population," *Scientific American*, 203: 194–204 (1960).

Denham, W. W., "Energy Relations and Some Basic Properties of Primate Social Organization," *American Anthropologist*, 73: 77–95 (1971).

DeSchlippe, P., *Shifting Cultivation in Africa: The Zande System of Agriculture*, Routledge and Kegan Paul, London, England, 1956.

Diener, P., "Ecology or Evolution?: the Hutterite Case," *American Ethnologist*, 1: 601–618 (1974).

Dimbleby, G. W., *Plants and Archaeology*, Humanities Press, New York, 1967.

Douglass, W., "Rural Exodus in Two Spanish Basque Villages: A Cultural Explanation," *American Anthropologist,* 73: 1100–1114 (1971).

———, Unpublished manuscript, Desert Research Institute, University of Nevada, Reno, n.d.

Drucker, P., *Indians of the Northwest Coast,* Doubleday, New York, 1963.

Duggan, A. J., "The Occurrence of Human Trypanosomiasis Among the Rukuba Tribe of Northern Nigeria," *Journal of Tropical Medicine,* 65: 151–63 (1962).

Dumond, D., "Population Growth and Political Centralization." In B. Spooner, ed., *Population Growth: Anthropological Implications,* MIT Press, Cambridge, Mass., 1972, pp. 286–310.

Dunn, F. L., "Epidemiological Factors: Health and Disease in Hunter-Gatherers." In R. Lee and I. DeVore, eds., *Man the Hunter,* Aldine, Chicago, 1968, pp. 221–28.

Easterlin, R., "On the Relation of Economic Factors to Recent and Projected Fertility Changes," *Demography,* 3: 131–53 (1966).

Eaton, J. W., and A. J. Mayer, "The Social Biology of Very High Fertility Among the Hutterites," *Human Biology,* 25: 206–63 (1953).

Edholm, O. G., *The Biology of Work,* McGraw-Hill, New York, 1967.

Eggan, F., *Social Organization of the Western Pueblos,* University of Chicago Press, Chicago, 1950.

Ehrlich, P. R., A. H. Ehrlich, and J. P. Holdren, *Human Ecology: Problems and Solutions,* W. H. Freeman, San Francisco, Calif., 1973.

Ehrlich, P. R. and P. H. Raven, "Butterflies and Plants: A Study in Coevolution," Evolution, 18: 586–608 (1964).

Elkin, A. P., *The Australian Aborigines,* 3rd ed., Natural History Press, Garden City, N.Y., 1964.

Elton, C., *Animal Ecology,* Macmillan, New York, 1927.

Engelmann, M.D., "The Role of Soil Arthopods in Community Energetics," *American Zoologist,* 8: 61–69 (1968).

Fenner, F., "The Effects of Changing Social Organisation on the Infectious Diseases of Man." In S. V. Boyden, ed., *The Impact of Civilisation on the Biology of Man,* University of Toronto Press, Toronto, Canada, and Australian National University Press, Canberra, Australia, 1970.

Fewkes, J. W., "The Tusayan Ritual: A Study of the Influence of Environment on Aboriginal Cults," *Annual Report of the Smithsonian Institution for 1895,* Washington, D.C., 1896, pp. 683–700.

Fishman, J., "A Systematization of the Whorfian Hypothesis," *Behavioral Science,* 5: 232–39 (1960).

Flannery, K. V., "The Ecology of Early Food Production in Mesopotamia," *Science,* 147: 1247–1256 (1965).

———, "Archaeological Systems Theory and Early Mesoamerica." In B. Meggers, ed., *Anthropological Archaeology on the Americas,* Anthropological Society of Washington, D.C., 1968, pp. 67–87.

———, "Effects of Early Domestication in Iran and the Near East." In P. J. Ucko and G. W. Dimbleby, eds., *The Domestication and Exploitation of Plants and Animals,* Aldine, Chicago, 1969, pp. 73–100.

———, "Summary Comments: Evolutionary Trends in Social Exchange and Interaction." In E. Wilmsen, ed., *Social Exchange and Interaction,* Anthropological Papers, No. 46, Museum of Anthropology, University of Michigan, Ann Arbor, 1972a, pp. 129–35.

———, "The Cultural Evolution of Civilizations," *Annual Review of Ecology and Systematics,* 3: 399–426 (1972b).

———, A. Kirkby, M. Kirby, and A. Williams, "Farming Systems and Political Growth in Ancient Oaxaca," *Science,* 158: 445–54 (1967).

Ford, R. I., "Barter, Gift, or Violence: An Analysis of Tewa Intertribal Exchange." In E. Wilmsen, ed., *Social Exchange and Interaction,* Anthropological Papers, No. 46, Museum of Anthropology, University of Michigan, Ann Arbor, 1972, pp. 21–45.

Forman, S., "Cognition and the Catch: The Location of Fishing Spots in a Brazilian Coastal Village," *Ethnology,* 6: 417–26 (1967).

Fowler, C. S., and J. Leland, "Some Northern Paiute Native Categories," *Ethnology,* 6: 381–404 (1967).

Fox, R., *Kinship and Marriage,* Penguin, Baltimore, 1967.

Frake, C. O., "The Diagnosis of Disease Among the Subanun of Mindanao," *American Anthropologist,* 63: 113–32 (1961).

———, "Cultural Ecology and Ethnography," *American Anthropologist,* 64(1, pt. 1): 53–59 (1962a).

———, "The Ethnographic Study of Cognitive Systems." In T. Gladwin and W. C. Sturtevant, eds., *Anthropology and Human Behavior,* Anthropological Society of Washington, 1962b, pp. 72–93.

———, "Further Discussion of Burling," *American Anthropologist,* 66: 119 (1964).

Freeman, J. D., *Iban Agriculture,* HMSO, London, England, 1955.

Freeman, J. L., "The Effects of Population Density on Humans." In J. T. Fawcett, ed., *Psychological Perspectives on Population,* Basic Books, New York, 1973, pp. 209–38.

Fried, M., *The Evolution of Political Society,* Random House, New York, 1967.

Gallucci, V. F., "On the Principles of Thermodynamics in Ecology," *Annual Review of Ecology and Systematics,* 4:329–357 (1973).

Garrod, D. A. E., and D. M. A. Bate, *The Stone Age of Mount Carmel, Vol. 1,* Oxford University Press, London, England, 1937.

Geertz, C., *Agricultural Involution: The Processes of Ecological Change in Indonesia,* University of California Press, Berkeley, 1963.

Gifford, D., "Ethnoarchaeological Observations on Post-Occupation Processes and the Problem of Archaeological Visibility," Paper given at the School of American Research Advanced Seminar on Ethnoarchaeology, Santa Fe, N. M., 1975.

Glacken, C., *Traces on the Rhodian Shore,* University of California Press, Berkeley, 1967.

Glob, P. V., *The Bog People,* Ballantine, New York, 1971.

Goldschmidt, W., "Social Organization in Native California and the Origin of the Clan," *American Anthropologist,* 50: 444–56 (1948).

Goodenough, W. H., "Componential Analysis and the Study of Meaning," *Language,* 32: 195–216 (1956).

———, "Cultural Anthropology and Linguistics," Report of the Seventh Annual

Round Table Meeting on Linguistics and Language Study, *Monograph Series on Languages and Linguistics,* No. 9, Washington, D.C., 1957, pp. 167–73.

Goodman, D., "The Theory of Diversity-Stability Relationships in Ecology," *The Quarterly Review of Biology,* 50: 237–66 (1975).

Greenwood, N. H. and J. M. B. Edwards, *Human Environments and Natural Systems: A Conflict of Dominion,* Wadsworth, Belmont, Calif., 1973.

Gross, D. R., "Protein Capture and Cultural Development in the Amazon Basin," *American Anthropologist,* 77: 526–549 (1975).

Hall, A. D., and R. E. Fagan, "Definition of System," *General Systems,* 1: 18–28 (1956).

Hall, E., *The Hidden Dimension,* Doubleday, New York, 1966.

Hall, R., "The Demographic Transition: Stage Four," *Current Anthropology,* 13: 212–15 (1972).

Hamburg, D. A., G. V. Coelho, and J. E. Adams, "Coping and Adaptation: Steps Toward a Synthesis of Biological and Social Perspectives." In D. A. Hamburg, G. V. Coelho, and J. E. Adams, eds., *Coping and Adaptation,* Basic Books, New York, 1974, pp. 403–40.

Hammel, E. A., ed., *Formal Semantic Analysis,* Special Publication of the American Anthropologist, 67(5, pt. 2), 1965.

Hanks, L., *Rice and Man: Agricultural Ecology in Southeast Asia,* Aldine, Chicago, 1972.

Hardesty, D. L., "The Niche Concept: Suggestions for Its Use in Studies of Human Ecology," *Human Ecology,* 3: 71–85 (1975).

Hardin, G., "The Competitive Exclusion Principle," *Science,* 131: 1291–1297 (1960).

Harlan, J. R., and D. Zohary, "Distribution of Wild Wheats and Barley," *Science,* 153: 1074–1080 (1966).

Harner, M., "Population Pressure and the Social Evolution of Agriculturalists," *Southwestern Journal of Anthropology,* 26: 67–86 (1970).

Harris, D., "The Prehistory of Tropical Agriculture: An Ethnoecological Model." In C. Renfrew, ed., *The Explanation of Culture Change: Models in Prehistory,* University of Pittsburgh Press, Pittsburgh, Pa., 1973, pp. 391–417.

Harris, M., *The Rise of Anthropological Theory*, Thomas Y. Crowell, New York, 1968.

———, *Culture, Man, and Nature*, Thomas Y. Crowell, New York, 1971.

———, *Cows, Pigs, Wars, and Witches: The Riddles of Culture*, Random House, New York, 1974a.

———, "Why a Perfect Knowledge of All the Rules One Must Know to Act Like a Native Cannot Lead to a Knowledge of How Natives Act," *Journal of Anthropological Research*, 30: 242–51 (1974b).

———, *Culture, People and Nature*, Thomas Y. Crowell, New York, 1975.

Harrison, G. A., and A. J. Boyce, "Migration, Exchange, and the Genetic Structure of Populations." In G. A. Harrison and A. J. Boyce, eds., *The Structure of Human Populations*, Oxford University Press, New York, 1972, pp. 128–45.

Harrison, G. A., J. S. Weiner, J. M. Tanner, and N. A. Barnicot, *Human Biology*, Oxford University Press, New York, 1964.

Hart, C. W. M., and A. R. Pilling, *The Tiwi of North Australia*, Holt, Rinehart and Winston, New York, 1960.

Hawley, A., "Ecology and Populations," *Science*, 179: 1196–1201 (1973).

Hayden, B., "Population Control Among Hunters/Gatherers," *World Archaeology*, 4: 205–21 (1972).

———, "The Carrying Capacity Dilemma." In A. C. Swedlund, ed., *Population Studies in Archaeology and Biological Anthropology: A Symposium*, Society for American Archaeology, Memoir 30, 1975, pp. 11–21.

Helbaek, H., "Paleo-Ethnobotany." In D. Brothwell and E. S. Higgs, eds., *Science in Archaeology, rev. ed.*, Thames and Hudson, London, England, 1969, pp. 206–14.

Hiatt, L. R., "Ownership and Use of Land Among the Australian Aborigines," In R. Lee and I. DeVore, eds., *Man the Hunter*, Aldine, Chicago, 1968, pp. 99–110.

Hodge, F. W., *Handbook of American Indians North of Mexico*, Smithsonian Institution, Bureau of American Ethnology, Bulletin 30, Washington, D.C., 1907.

Hole, F., and R. F. Heizer, *An Introduction to Prehistoric Archaeology*, 3rd ed., Holt, Rinehart and Winston, New York, 1973.

Holling, C. S., "Resilience and Stability of Ecological Systems," *Annual Review of Ecology and Systematics,* 4: 1–23 (1973).

Holling, C. S., and M. A. Goldberg, "Ecology and Planning," *Journal of the American Institute of Planners,* 37: 221–30 (1971).

Holmberg, A. R., *Nomads of the Long Bow,* Natural History Press, Garden City, New York, 1969.

Holmes, W. H., *Handbook of Aboriginal American Antiquities,* Smithsonian Institution, Bureau of American Ethnology, Bulletin 60, Part 1, Washington, D.C., 1919.

Howell, N., "The Feasibility of Demographic Studies in 'Anthropological' Populations." In M. H. Crawford and P. L. Workman, eds., *Methods and Theories of Anthropological Genetics,* University of New Mexico Press, Albuquerque, 1973, pp. 249–62.

Huntington, E., *Mainsprings of Civilization,* John Wiley & Sons, New York, 1945.

Hutchinson, G. E., *The Ecological Theater and the Evolutionary Play,* Yale University Press, New Haven, Conn., 1965.

Hymes, D. H., "Discussion of Burling's Paper," *American Anthropologist,* 66: 116–19 (1964).

Iltis, H. H., O. L. Loucks, and P. Andrews, "Criteria for an Optimum Human Environment," *Bulletin of the Atomic Scientists,* 26: 2–6 (1970).

Janzen, D. H., "Tropical Agroecosystems," *Science,* 182: 1212–1219 (1973).

Johnson, A., "Ethnoecology and Planting Practices in a Swidden Agricultural System," *American Ethnologist,* 1: 87–101 (1974).

Johnson, G., "A Test of the Utility of Central Place Theory in Archaeology." In P. J. Ucko, R. Tringham, and G. W. Dimbleby, eds., *Man, Settlement, and Urbanism,* Gerald Duckworth, London, England, and Schenkman, Cambridge, Mass., 1972, pp. 769–85.

Jolly, C., "The Seed-Eaters: A New Model of Hominid Differentiation Based on a Baboon Analogy," *Man,* 5: 5–26 (1970).

Kaplan, D., and R. Manners, *Culture Theory,* Prentice-Hall, Englewood Cliffs, N.J., 1972.

Katz, S., "Biological Factors in Population Control." In B. Spooner, ed., *Population Growth: Anthropological Implications,* MIT Press, Cambridge, Mass., 1972, pp. 351–69.

Kay, P., "Some Theoretical Implications of Ethnographic Semantics." In *Current Directions in Anthropology,* Bulletins of the American Anthropological Association, 3(3, pt. 2), Washington, D.C., 1970, pp. 19–31.

———, "Comments on Colby." In S. A. Tylor, ed., *Cognitive Anthropology,* Holt, Rinehart, and Winston, New York, 1969, pp. 78–90.

Keesing, R. M., "Kwara?ae Ethnoglottochronology: Procedures Used by Malaita Cannibals for Determining Percentages of Shared Cognates," *American Anthropologist,* 75: 1282–1289 (1973).

Klein, R. G., *Man and Culture in the Late Pleistocene,* Chandler, San Francisco, 1969.

Kormondy, E. J., *Concepts of Ecology,* Prentice-Hall, Englewood Cliffs, N.J., 1969.

Kortlandt, A., "New Perspectives on Ape and Human Evolution," *Current Anthropology,* 15: 427–48 (1974).

Kozlowsky, D. C., "Critical Evaluation of Trophic Level Concepts," *Ecology,* 49: 48–60 (1968).

Kroeber, A. L., *Cultural and Natural Areas of Native North America,* University of California Press, Berkeley, 1939.

Krumbein, W. C., "Sampling in Paleontology." In B. Kummel and D. Raup, eds., *Handbook of Paleontological Techniques,* W. H. Freeman, San Francisco, 1965, pp. 137–49.

Kummer, H., *Primate Societies,* Aldine, Chicago, 1971.

Kunstadter, P., "Demography, Ecology, Social Structure, and Settlement Patterns." In G. A. Harrison and A. J. Boyce, eds., *The Structure of Human Populations,* Oxford University Press, New York, 1972, pp. 313–51.

Lasker, G., "Human Biological Adaptability," *Science,* 166: 1480–1486 (1969).

Lathrap, D., *The Upper Amazon,* Praeger, New York, 1970.

Lee, R., "What Hunters Do for a Living, or, How to Make Out on Scarce Resources." In R. Lee and I. DeVore, eds., *Man the Hunter,* Aldine, Chicago, 1968. pp. 30–48.

———, "!Kung Bushman Subsistence: An Input-Output Analysis." In D. Damas, ed., *Contributions to Anthropology: Ecological Essays*, Natural Museums of Canada, Bulletin 230, Ottawa, 1969, pp. 73–94.

———, "!Kung Spatial Organization: An Ecological and Historical Perspective," *Human Ecology*, 1: 125–47 (1972a).

———, "Population Growth and the Beginnings of Sedentary Life Among the !Kung Bushmen." In B. Spooner, ed., *Population Growth: Anthropological Implications*, MIT Press, Cambridge, Mass., 1972b, pp. 329–42.

Lee, R., and I. DeVore, "Problems in the Study of Hunters and Gatherers." In R. Lee and I. DeVore, eds., *Man the Hunter*, Aldine, Chicago, 1968, pp. 3–12.

Levi-Strauss, C., *The Savage Mind*, University of Chicago Press, Chicago, 1966.

Levins, R., *Evolution in Changing Environments*, Princeton University Press, Princeton, N.J., 1968.

Lindeman, R. L., "The Trophic-Dynamic Aspect of Ecology," *Ecology*, 23: 399–418 (1942).

Lord, E., *Comstock Mining and Miners*, U.S. Geological Survey, Government Printing Office, Washington, D.C., 1883.

Lotka, A. J., *Elements of Physical Biology*, Williams and Wilkins, Baltimore, 1925.

MacArthur, R., "On the Relative Abundance of Species," *American Naturalist*, 94: 25–36 (1960).

McLaren, I. A., "Introduction." In I. A. McLaren, ed., *Natural Regulation of Animal Populations*, Atherton, New York, 1971, pp. 1–21.

Malkin, B., "Sumu Ethnozoology: Herpetological Knowledge," *Davidson Journal of Anthropology*, 2: 165–80 (1956).

———, "Cora Ethnozoology, Herpetological Knowledge: A Bio-Ecological and Cross Cultural Approach," *Anthropological Quarterly*, 31: 73–90 (1958).

———, *Seri Ethnozoology*, Idaho State College Museum, Occasional Papers, No. 7, 1962.

Margalef, D. R., *Perspectives in Ecological Theory*, University of Chicago Press, Chicago, 1968.

Mason, O. T., "Influence of Environment upon Human Industries or Arts," Annual Report of the Smithsonian Institution for 1895, Washington, D.C., 1896, pp. 639–65.

May, J., "Influence of Environmental Transformation in Changing the Map of Disease." In M. T. Farvar and J. P. Milton, eds., *The Careless Technology: Ecology and International Development,* Natural History Press, Garden City, N.Y., 1972, pp. 19–34.

May, R. M., "Ecosystem Patterns in Randomly Fluctuating Environments," *Progress in Theoretical Biology,* 3, pp. 1–50 (1974).

Mayr, E., *Populations, Species, and Evolution,* Belknap Press, Harvard University, Cambridge, Mass., 1970.

Mazess, R. B., "Human Adaptation to High Altitude." In A. Damon, ed., *Physiological Anthropology,* Oxford University Press, New York, 1975, pp. 167–209.

Meggers, B., "Environmental Limitations on the Development of Culture," *American Anthropologist,* 56: 801–24 (1954).

———, *Amazonia,* Aldine, Chicago, 1971.

———, *Prehistoric America,* Aldine, Chicago, 1972.

———, "Application of the Biological Model of Diversification to Cultural Distributions in Tropical Lowland South America," *Biotropica,* 7: 141–61 (1975).

Meggitt, M. J., "The Association Between Australian Aborigines and Dingoes." In A. Leeds and A. P. Vayda, eds., *Man, Culture, and Animals,* American Association for the Advancement of Science, Publication No. 78, 1965, pp. 7–26.

———, "System and Subsystem: The Te Exchange Cycle Among the Mae Enga," *Human Ecology,* 1: 111–23 (1972).

Messenger, J., *Inis Beag, Isle of Ireland,* Holt, Rinehart and Winston, New York, 1969.

Metzger, D., and G. Williams, "Some Procedures and Results in the Study of Native Categories: Tzeltal 'Firewood,' " *American Anthropologist,* 68: 389–407 (1966).

Miller, G. T., *Living in the Environment,* Wadsworth, Belmont, Calif. 1975.

Montagu, A., ed., *Man and Aggression,* Oxford University Press, New York, 1968.

Moore, J. A., A. C. Swedlund, and G. J. Armelagos, "The Use of Life Tables in Paleodemography." In A. C. Swedlund, ed., *Population Studies in Archaeology and Biological Anthropology: A Symposium,* Memoirs of the Society for American Archaeology, No. 30, 1975, pp. 57–70.

Moran, J., M. D. Morgan, and J. H. Wiersma, *An Introduction to Environmental Sciences,* Little, Brown, Boston, 1973.

Moriber, G., *Environmental Science,* Allyn and Bacon, Inc., Boston, 1974.

Morrill, W. T., "Ethnoicthyology of the Cha-Cha," *Ethnology,* 6: 405–16 (1967).

Morrison, J. P. E., "Preliminary Report on Mollusks Found in the Shell Mounds of the Pickwick Landing Basin in the Tennessee River Valley." In W. S. Webb and D. L. DeJarnette, *An Archaeological Survey of Pickwick Basin in the Adjacent Portions of the States of Alabama, Mississippi and Tennessee,* Smithsonian Institution, Bureau of American Ethnology, Bulletin 129, 1942, pp. 337–92.

Mueller, J. W., *The Use of Sampling in Archaeological Survey,* Memoirs of the Society for American Archaeology, No. 28, 1974.

Murdoch, W. W. and A. Oaten, "Predation and Population Stability," *Advances in Ecological Research* 9: 1–31 (1975).

Nag, M., *Factors Affecting Human Fertility in Non-Industrial Societies: A Cross-Cultural Study,* Yale University Publications in Anthropology, No. 66, 1962.

———, "Anthropology and Population: Problems and Perspectives," *Population Studies,* 27: 59–68 (1973).

Naroll, R., "A Preliminary Index of Social Development," *American Anthropologist,* 58: 687–715 (1956).

Netting, R. M., "Fighting, Forest, and the Fly: Some Demographic Regulators Among the Kofyar," *Journal of Anthropological Research,* 29: 164–79 (1973).

———, "Agrarian Ecology," *Annual Review of Anthropology,* 3: 21–56 (1974).

Newman, L. F., *Birth Control: An Anthropological View,* Addison-Wesley Module in Anthropology, No. 27, Addison-Wesley, Reading, Mass., 1972.

Nietschmann, B., *Between Land and Water, the Subsistence Ecology of the Miskito Indians, Eastern Nicaragua,* Seminar Press, New York, 1973.

Odum, E. P., *Fundamentals of Ecology, 3rd ed.*, W. Saunders, Philadelphia, 1971.

Odum, H. T., *Environment, Power, and Society,* John Wiley & Sons, New York, 1971.

Oliver, S. C., *Ecology and Cultural Continuity as Contributing Factors in the Social Organizations of the Plains Indians,* University of California Publications in American Archaeology and Ethnology, 48(1), 1962.

Parsons, J. R., "The Archaeological Significance of Mahamaes Cultivation on the Coast of Peru," *American Antiquity,* 33:80–85 (1968).

Perchonock, N., and O. Werner, "Navaho Systems of Classification: Some Implications for Ethnoscience," *Ethnology,* 8: 229–42 (1969).

Peterson, W., "A Demographer's View of Prehistoric Demography," *Current Anthropology,* 16: 227–45 (1975).

Phillipson, J., *Ecological Energetics,* Studies in Biology, No. 1, Edward Arnold, London, England, 1966.

Piddocke, S., "The Potlatch System of the Southern Kwakiutl: A New Perspective," *Southwestern Journal of Anthropology,* 21: 244–64 (1965).

Pike, K. L., *Language in Relation to a Unified Theory of the Structure of Human Behavior,* Parts I (1954), II (1955), and III (1960), Summer Institute of Linguistics, Glendale, Calif., 1954–1960.

Polanyi, K., "The Economy as Instituted Process." In K. Polanyi, C. M. Arensberg, and H. W. Pearson, eds., *Trade and Market in the Early Empires,* Free Press, New York, 1957, pp. 243–70.

Polgar, S., "Population History and Population Policies from an Anthropological Perspective," *Current Anthropology,* 13: 203–11 (1972).

Pospisil, L., *Kapauku Papuan Economy,* Yale University Publications in Anthropology, No. 67, 1963.

Pressat, R., *Demographic Analysis,* Translated by Judah Matras, Aldine, Chicago, 1972.

Preston, R. J., *North American Trees, rev. ed.*, MIT Press, Cambridge, Mass., 1966.

Puleston, D. E. and O. S. Puleston, "An Ecological Approach to the Origins of Maya Civilization, *Archaeology,*" 124: 330–337 (1971).

Radcliffe-Brown, A. R., *The Andaman Islanders,* Free Press, New York, 1948.

Rappaport, R. A., "Aspects of Man's Influence on Island Ecosystems: Alteration and Control." In F. R. Fosberg, ed., *Man's Place in the Island Ecosystem,* Bishop Museum Press, Honolulu, 1963, pp. 155–74.

———, *Pigs for the Ancestors,* Yale University Press, New Haven, Conn., 1968.

———, "The Sacred in Human Evolution," *Annual Review of Ecology and Systematics,* 2: 25–44 (1971).

Rapport, D. J. and J. E. Turner, "Feeding rates and Population Growth," *Ecology,* 56: 942–949 (1975).

Rathje, W., "The Origin and Development of Lowland Classic Maya Civilization," *American Antiquity,* 36: 275–85 (1971).

Renfrew, C., *The Emergence of Civilisation, the Cyclades and the Aegean in the Third Millennium* B.C., Methuen, London, England, 1972.

Richard, A., *Land, Labour, and Diet in Northern Rhodesia,* Oxford University Press, London, England, 1939.

Richmond, R. C., "Non-Darwinian Evolution: A Critique," *Nature,* 225: 1025–1028 (1970).

Robbins, W. W., J. P. Harrington, and B. Freire-Marreco, *Ethnobotany of the Tewa Indians,* Smithsonian Institution, Bureau of American Ethnology, Bulletin 55, 1916.

Roberts, D. F., "Genetic Effects of Population Size Reduction," *Nature,* 220: 1084–1088 (1968).

Rogers, E. S., "The Mistassini Cree." In M. G. Bicchieri, ed., *Hunters and Gatherers Today,* Holt, Rinehart and Winston, New York, 1972, pp. 90–137.

Rosenzweig, M. L., "Net Primary Productivity of Terrestrial Communities; Predictions from Climatological Data," *American Naturalist,* 102:67–74 (1968).

Roughgarden, J., "Evolution of Niche Width," *American Naturalist,* 106: 683–718 (1972).

Ruyle, E. E., "Genetic and Cultural Pools: Some Suggestions for a Unified Theory of Biocultural Evolution," *Human Ecology,* 1: 201–215 (1973).

Sahlins, M., "Land Use and the Extended Family in Moala, Fiji," *American Anthropologist,* 59: 449–62 (1957).

———, *Social Stratification in Polynesia,* University of Washington Press, Seattle, 1958.

———, "The Segmentary Lineage: An Organization of Predatory Expansion," *American Anthropologist,* 63: 322–43 (1961).

———, *Tribesmen,* Prentice-Hall, Englewood Cliffs, N.J., 1966.

Salzano, F., and N. Freire-Maia, *Problems in Human Biology: A Study of Brazilian Populations,* Wayne State University Press, Detroit, Mich., 1970.

Sanders, W., and B. Price, *Mesoamerica, the Evolution of a Civilization,* Random House, New York, 1968.

Sapir, E., "Language and Environment," *American Anthropologist,* 14: 226–42 (1912).

———, "The Psychological Reality of Phonemes" (Original title in French, 1933). In D. G. Mandelbaum, ed., *Selected writings of Edward Sapir in Language, Culture and Personality,* University of California Press, Berkeley, 1958, pp. 46–60.

Schneider, H. K., *Economic Man,* Free Press, New York, 1974.

Schofield, F. D., "Some Relations Between Social Isolation and Specific Communicable Diseases," *The American Journal of Tropical Medicine and Hygiene,* 19: 167–69 (1970).

Schwartz, D. W., "Demographic Changes in the Early Periods of Cohonina Prehistory." In G. R. Willey, ed., *Prehistoric Settlement Patterns in the New World,* Viking Fund Publications in Anthropology, No. 23, 1956, pp. 26–31.

Schoener, T. W., "Theory of Feeding Strategies," *Annual Review of Ecology and Systematics,* 2: 369–404 (1971).

Scott, D., "The Epidemiology of Human Trypanosomiasis in Ashanti, Ghana," *Journal of Tropical Medicine and Hygiene,* 60: 202–14, 257–74, 303–15 (1957).

Service, E., *Primitive Social Organization, 2nd ed.,* Random House, New York, 1971a.

———, *Cultural Evolutionism, Theory in Practice,* Holt, Rinehart and Winston, New York, 1971b.

Shackley, M. L., *Archaeological Sediments,* Halstead, London, England, 1975.

Shelford, V. E., *Animal Communities in Temperate America,* University of Chicago Press, Chicago, 1913.

Slobodkin, L. B., "Toward a Predictive Theory of Evolution." In R. C. Lewontin, ed., *Population Biology and Evolution,* Syracuse University Press, Syracuse, N.Y., 1968, pp. 187–205.

———, and A. Rapoport, "An Optimal Strategy of Evolution," *The Quarterly Review of Biology,* 49: 181–200 (1974).

Smith, C., "Economics of Marketing Systems: Models from Economic Geography," *Annual Review of Anthropology,* 3: 167–201 (1974).

Smith, R. L., ed., *The Ecology of Man,* Harper and Row, New York, 1972.

———, *Ecology and Field Biology,* 2nd ed., Harper & Row, New York, 1974.

Society of Economic and Mineralogic Petrologists, *Recognition of Ancient Sedimentary Environments,* Special Publication No. 16, 1972.

Sommer, R., *Personal Space,* Prentice-Hall, Englewood Cliffs, N.J., 1969.

Sparks, B. W., "Non-Marine Mollusca and Archaeology." In D. Brothwell and E. S. Higgs, eds., *Science in Archaeology,* Basic Books, New York, 1963, pp. 313–23.

Spooner, B., "Introduction." In B. Spooner, ed., *Population Growth: Anthropological Implications,* MIT Press, Cambridge, Mass., 1972, pp. xv–xxvii.

Stevenson, R. F., *Population and Political Systems in Tropical Africa,* Columbia University Press, New York, 1968.

Steward, J. H., "Irrigation Without Agriculture," *Papers of the Michigan Academy of Science, Arts and Letters,* 12: 149–56 (1930).

———, *Basin-Plateau Aboriginal Sociopolitical Groups,* Smithsonian Institution, Bureau of American Ethnology, Bulletin 120, 1938.

———, *Theory of Culture Change,* University of Illinois Press, Urbana, 1955.

Stewart, O. C., "Fire as the First Great Force Employed By Man." In W. L. Thomas, ed., *Man's Role in Changing the Face of the Earth,* University of Chicago Press, Chicago, 1956, pp. 115–33.

Street, J., "An Evaluation of the Concept of Carrying Capacity," *Professional Geographer,* 21:104–107 (1969).

Streuver, S., "Flotation Techniques for the Recovery of Small-Scale Archaeological Remains," *American Antiquity,* 33: 353–62 (1968a).

———, "Woodland Subsistence-Settlement Systems in the Lower Illinois Valley." In S. R. Binford and L. R. Binford, eds., *New Perspectives in Archaeology,* Aldine, Chicago, 1968b, pp. 285–312.

Sturtevant, W. C., "Studies in Ethnoscience." In A. K. Romney and R. G. D'Andrade, eds., *Transcultural Studies in Cognition,* Special Publication of the *American Anthropologist,* 66(3, pt. 2), 1964, pp. 99–131.

Swedlund, A. S., "The Use of Ecological Hypotheses in Australopithecine Taxonomy," *American Anthropologist,* 75: 515–29 (1974).

Sweet, L., "Camel Raiding of North Arabian Bedouin: A Mechanism of Ecological Adaptation," *American Anthropologist,* 67: 1132–1150 (1965).

Teitelbaum, M. S., "Factors Associated with the Sex Ratio in Human Populations." In G. A. Harrison and A. J. Boyce, eds., *The Structure of Human Populations,* Oxford University Press, New York, 1972, pp. 90–109.

Thomas, D. H., "Western Shoshoni Ecology: Settlement Patterns and Beyond." In D. D. Fowler, ed., *Great Basin Cultural Ecology: A Symposium,* Desert Research Institute Publications in the Social Sciences, No. 8, 1972, pp. 135–53.

Thomas, R. B., *Human Adaptation to a High Andean Energy Flow System,* Occasional Papers in Anthropology, No. 7, The Pennsylvania State University, University Park, 1973.

———, "The Ecology of Work." In A. Damon, ed., *Physiological Anthropology,* Oxford University Press, New York, 1975, pp. 59–79.

Trigger, B., "The Determinants of Settlement Patterns." In K. C. Chang, ed., *Settlement Archaeology,* National Press, Palo Alto, Calif., 1968, pp. 53–78.

Tyler, S. A., "Introduction." In S. A. Tyler, ed., *Cognitive Anthropology,* Holt, Rinehart and Winston, New York, 1969, pp. 1–27.

Vayda, A. P., "Warfare in Ecological Perspective," *Annual Review of Ecology and Systematics,* 5: 21–31 (1974).

———, and B. McCay, "New Directions in Ecology and Ecological Anthropology," *Annual Review of Anthropology,* 4: 293–306 (1975).

———, and R. Rappaport, "Ecology, Cultural and Noncultural." In J. Clifton, ed., *Introduction to Cultural Anthropology,* Houghton Mifflin, Boston, 1968, pp. 477–97.

Vigas, V., "Kuru in New Guinea: Discovery and Epidemiology," *The American Journal of Tropical Medicine and Hygiene,* 19: 130–32 (1970).

———, and D. C. Gajdusek, "Degenerative Disease of the Central Nervous System in New Guinea, the Endemic Occurrence of 'Kuru' in the Native Population," *New England Journal of Medicine,* 257: 974–78 (1957).

de Vos, A., and H. Mosby, "Habitat Analysis and Evaluation." In R. Giles, ed., *Wildlife Management Techniques,* Wildlife Society, Washington, D.C., 1969, pp. 135–72.

Wagner, G., *The Bantu of North Kavirondo, Vol. 2: Economic Life,* Oxford University Press, New York, 1954.

Wallace, A. F. C., "Mental Illness, Biology, and Culture." In F. L. K. Hsu, ed., *Psychological Anthropology: Approaches to Culture and Personality,* Dorsey Press, Homewood, Ill., 1961, pp. 255–95.

———, and J. Atkins, "The Meaning of Kinship Terms," *American Anthropologist,* 62: 58–80 (1960).

Walters, C. J., "System Ecology: The Systems Approach and Mathematical Models in Ecology." In E. Odum, *Fundamentals of Ecology, 3rd ed.,* W. Saunders, Philadelphia, 1971, pp. 276–92.

Wedel, W., *Environment and Native Subsistence Economies in the Central Great Plains,* Smithsonian Institution, Miscellaneous Collections, 100 (3), 1941.

Weins, J. A., "On Group Selection and Wynne-Edwards Hypothesis," *American Scientist,* 54: 273–87 (1966)

Weiner, J., "Tropical Ecology and Population Structure." In G. A. Harrison and A. J. Boyce, eds., *The Structure of Human Populations,* Oxford University Press, New York, 1972, pp. 393–410.

Weiss, K., *Demographic Models in Anthropology*, Memoirs of the Society for American Archaeology, No. 27, 1973.

Wilkinson, R. G., *Poverty and Progress: An Ecological Perspective on Economic Development*, Praeger, New York, 1973.

Willey, G. R., *Prehistoric Settlement Patterns in the Viru Valley, Peru*, Smithsonian Institution, Bureau of American Ethnology, Bulletin 155, 1953.

Williams, B. J., *A Model of Band Society*, Memoirs of the Society for American Archaeology, No. 29, 1974.

Williams, G. C., ed., *Group Selection*, Aldine, Chicago, 1971.

Williams, G. C., *Adaptation and Natural Selection: A Critique of Some Current Evolutionary Thought*, Princeton University Press, Princeton, 1966.

Wilmsen, E., "Introduction: The Study of Exchange as Social Interaction." In E. Wilmsen, ed., *Social Exchange and Interaction*, Anthropological Papers, No. 46, Museum of Anthropology, University of Michigan, Ann Arbor, 1972, pp. 1–4.

Wissler, C., *The Relation of Nature to Man in Aboriginal America*, Oxford University Press, New York, 1926.

Wittfogel, K., *Oriental Despotism*, University of Washington Press, Seattle, 1957.

Wolpoff, M., "Interstitial Wear," *American Journal of Physical Anthropology*, 34: 205–25 (1971).

Wright, G. A., *Archaeology and Trade*, Addison-Wesley Module in Anthropology, No. 49, Addison-Wesley, Reading, Mass., 1974.

Wyman, L. C., and F. L. Bailey, *Navaho Indian Ethnoentomology*, University of New Mexico Publications in Anthropology, No. 12, Albuquerque, N.M., 1964.

Wynne-Edwards, V. C., *Animal Dispersion in Relation to Social Behavior*, Hafner, New York, 1962.

Yengoyan, A. A., "Demographic and Ecological Influences on Aboriginal Australian Marriage Sections." In R. Lee and I. DeVore, eds., *Man the Hunter*, Aldine, Chicago, 1968, pp. 185–99.

Zubrow, E. B. W., *Prehistoric Carrying Capacity: A Model*, Cummings, Menlo Park, Calif., 1975.

GLOSSARY

ACCLIMATIZATION A reversible physiological response to environmental stress.

ACTUAL EVAPOTRANSPIRATION The amount of water that evaporates from the surface in the form of runoff or percolation plus the amount of water that transpires from the stomata of plants.

ADAPTABILITY Resilience.

ADAPTATION A beneficial response to environmental stress.

ADAPTIVE RADIATION Explosive evolution from an unspecialized ancestor into a diverse variety of descendants each specialized for a specific environment.

AGE PYRAMID A graphical representation of a population's age and sex structure.

AGRICULTURE A farming method based upon drastic alteration of the natural ecological system and the creation of an artificial ecological system based upon domesticated plants and imported nutrients.

ALBEDO The energy reflecting power of the earth.

AMINO ACIDS The complex organic molecules making up proteins.

ANTIBIOSIS Interactions between populations that result in definite harm to one but not to the other.

ASSIMILATION EFFICIENCY The extent to which food passing from one trophic level to another is actually metabolized and used.

ATMOSPHERE The thin layer of gases (air) surrounding the earth.

AUSTRALOPITHECINES Early African hominids, living about 5,000,000 to 750,000 years ago.

AUTECOLOGY An approach to ecological study focusing upon the environmental relationships of a population.

AUTOTROPH An organism capable of photosynthesis, a producer; mainly green plants.

BILATERAL AFFILIATION Kinship ties traced through both the father and mother and their relatives.

BIOGEOCHEMICAL CYCLE A material cycle having biological, geological, and chemical stages.

BIOSPHERE The thin layer from about 6 feet above the earth to 6 feet below within which most living organisms exist.

"BOOM-BUST" POPULATION GROWTH A model of population growth resembling an exponential curve to a point, then dropping drastically to a much smaller size; a J-shaped curve.

CALORIE OR GRAM CALORIE The amount of heat required to increase the temperature of one gram of water from 3.5 to 4.5 degrees Celsius.

CARGO SYSTEM A Mesoamerican institution in which prestige positions are rotated among members of the community. Because of the expense of occupying the positions, the cargo system is a way of redistributing wealth within the community.

CARNIVORE A consumer organism that feeds upon animals.

CARRYING CAPACITY The theoretical limit to which a population can grow and still be permanently supported by the environment.

CHEMICAL ENERGY Energy bound in complex molecules, such as sugars, and released by oxidation.

CHIEFDOM A ranked society with status dependent upon kinship.

CLAN A group of kin that trace descent from a common, mythical ancestor.

COEVOLUTION Mutual evolutionary change in two or more groups of organisms with close ecological relationships.

COGNATIC DESCENT Ancestry through either the male or female line.

COMMENSALISM Interaction between populations that is beneficial to one and neither good nor harmful to the other.

COMPETITIVE FEASTING A ritual, such as the potlach, in which feasts are given by social climbers for the purpose of gaining higher status.

COMPONENTIAL ANALYSIS A method used by ethnoscience to analyze the semantic elements of a paradigm; each member of the paradigm is defined by the intersection of components of meaning.

COMPOSITE BAND A group of unrelated nuclear families linked by ritual or fictive kinship ties or a multifamily group linked by bilateral descent.

CONDUCTION Energy exchange between objects in direct contact.

CONNUBIUM The largest group within which marriage usually takes place.

CONTRAST (LINGUISTIC) A principle of semantic relationship used in taxonomies wherein terms at any single level can be viewed as mutually exclusive in meaning and use; for example, a "pine" is not a "fir."

CONVECTION Energy exchange between an object and a moving fluid.

COOPERATION Interaction between populations in which both are benefited by the relationship.

COPROLITES The preserved fecal matter of animals.

CORVEE LABOR Organized public labor, used in building public monuments, dams, and so forth.

COST-YIELD RATIO The calories of food yield for each calorie of work expended in its production or exploitation.

CROSS-COUSIN MARRIAGE Marriage between the children of siblings of the opposite sex.

CRUDE FERTILITY RATE The number of births per 1000 population during a calendar year.

CRUDE MORTALITY RATE The number of deaths per 1000 population during a calendar year.

CULTURAL ECOLOGY The study of relationships between culture and environment.

CULTURE The pattern of behavior learned by humans as members of a social group. It is symbolic, shared by members of a social group, and passed on from one generation to another.

CULTURE AREA A geographical area within which human groups have similar cultures.

CULTURE CORE As defined by Julian Steward, those features of a culture that are most closely related to subsistence.

DARWINIAN EVOLUTION Evolutionary change by the process of natural selection.

DARWINIAN FITNESS An index of relative reproductive success that is often used to measure adaptive success.

DEMOGRAPHIC TRANSITION THEORY A model of human population growth based upon data from Western Europe, approximating the logistic or S-shaped growth pattern.

DEMOGRAPHY The description and prediction of population characteristics.

DENSITY-DEPENDENT POPULATION REGULATION The regulation of population growth by internal processes controlled by population size or density.

DENSITY-INDEPENDENT POPULATION REGULATION The regulation of population growth by external ("environmental") processes having nothing to do with population size or density.

DETRITOVORES Organisms that feed upon dead matter.

DETRITUS CHAIN A food chain that starts with the dead matter sluffed off by grazing chains.

DIRECTIONAL SELECTION Selection process that increases the frequency of novel genotypes that are advantageous.

DIVERSIFYING EVOLUTION (OR CLADOGENESIS) Evolutionary change in the direction of increased adaptation to specific environments. Contrasts with progressive evolution that increases general adaptability.

DIVERSIFYING SELECTION Selection process that increases the frequency of several genotypes simultaneously.

DOMAIN (LINGUISTIC) A sphere of meaning; a major category or a classification system defined in terms of a specific culture, for example, "plants" in English.

ECOGEOGRAPHICAL RULE A statement of a relationship between the physical characteristics of an organism and a geographically variable feature of the environment.

ECOLOGICAL ANTHROPOLOGY The study of relationships between humans and their environment.

ECOLOGICAL COMMUNITY An interacting group of plants and/or animals; organisms living together.

ECOLOGICAL ENERGETICS The study of energy pathways, exchanges, and transformations in an ecological system.

ECOLOGICAL NICHE A distinctive life style of an organism, usually defined by a unique use of resources.

ECOLOGICAL OVERLAP Marginal habitats occupied in common by competing populations.

ECOLOGICAL SUCCESSION A series of changes in ecological communities in which opportunistic populations are replaced by equilibrium populations.

ECOLOGICAL SYSTEM (OR ECOSYSTEM) An interacting group of plants and animals, along with their nonliving environment.

ECOLOGY That branch of science concerned with the relationships between organisms and their environment.

EFFICIENCY INDEX The ratio of energy output to energy input, times 100.

ELECTROMAGNETIC ENERGY Energy in the forms of waves that can be radiated in a vacuum.

EMICS The significant features of the behavioral and structural universe of a particular culture described from the point of view of that culture. (Also, see Etics.)

ENDOGAMY A rule that requires members of the same group to marry each other.

ENERGY MAXIMIZER A feeding strategy in which the organism maximizes food yield.

ENTROPY Degraded energy that cannot be used for work; a measure of energy disorder or randomness.

ENVIRONMENT Surroundings that enter into relationships with a thing under consideration.

ENVIRONMENTAL DETERMINISM The idea that environment directly causes human behavior and biology.

ENVIRONMENTAL GRADIENTS A continuous variable in the environment, such as mean annual rainfall, soil temperature, actual evapotranspiration, and so forth.

EQUILIBRIUM A static concept in which a system is viewed as being in perfect balance with its environment.

EQUILIBRIUM ORGANISMS Organisms with specialized physiology and behavior suitable for living in a stable environment.

ETHNOECOLOGY A distinctive approach to human ecology that focuses on the conceptions of ecological relationships held by a people or culture.

ETHNOGRAPHIC ANALOGY Observing living peoples and cultures to draw conclusions about the past.

ETHNOSCIENCE The study of systems of knowledge developed by a given culture to classify the objects, activities, and events of its universe.

ETHNOSYSTEMATICS The study of native schemes for classifying the objects of their natural universe.

ETICS Aspects of the behavioral and structural universe of a culture described in terms of culture-free features or features derived from comparisons of many cultures. (Also, see emics.)

EVAPORATION Energy exchange caused by conversion from a liquid to a gas or vice versa.

EVOLUTION Change.

EXOGAMY A rule that prohibits members of the same group from marrying each other.

EXTERNALITY Any indirect cost or benefit existing outside the marketplace and not included in the price of a good or service.

FALLOW A stage in a farming cycle when a field is not being farmed.

FANDANGO A Western Shoshoni gathering taking place after the pine nut harvest.

FECUNDITY The reproductive potential of an organism; sometimes used to mean actual reproduction.

FEEDBACK A response that is used to control or influence a future response.

FEEDING Energy exchange between organisms by means of biomass.

FERTILITY RATE The number of births taking place within a given group over a given period of time.

FERTILITY RATIO The number of persons in the youngest age class divided by the number of women of child-bearing age multiplied by 100; also called the "child-woman ratio."

FIRST LAW OF THERMODYNAMICS The Law of Energy Conservation; states that energy can be neither created nor destroyed but can be transformed.

FOOD CHAIN A series of organisms that feed upon each other; an energy pathway.

FOOD WEB A set of interrelated food chains or energy pathways.

GAME THEORY A way of determining the conditions under which gains are maximized and losses minimized.

GASEOUS CYCLES The circular pathway taken by materials that exist at some point in a gaseous state, for example, carbon.

GENERAL FERTILITY RATE The number of births per 1000 women of child-bearing age.

GENOTYPE Genetic instructions for the production of a given physical characteristic.

GRAND HYSTERIA A form of hysteria common in nineteenth century East European cities and possibly caused by calcium deficiency in the diet.

GRAZING CHAIN A food chain that starts with green plants or other producers.

GROSS PRIMARY PRODUCTIVITY Primary productivity uncorrected for energy loss from respiration.

GROSS REPRODUCTIVE RATE The number of female children born per 1000 women of child-bearing age.

GROUP SELECTION The idea that differential survival is applicable to groups as well as to individuals.

HABITAT The geographical area occupied by a population; not to be confused with the ecological niche, which is how the population makes a living in its habitat.

HEAT ENERGY Energy caused by the random movement of molecules.

HERBIVORE An organism that feeds upon producers.

HETEROTROPH An organism that is not capable of photosynthesis and must feed upon other organisms.

HOMEOSTASIS The dynamic idea that systems are continually changing but tend to change within limits.

HUMOUR THEORY Hippocrates's idea that four substances, or "humours," controlled human behavior and biology, widely accepted until the nineteenth century and associated with environmental determinism.

HYDROSPHERE That part of the earth dominated by water.

HYPERCOHERENCE A pathological condition in systems resulting from too many interconnections of its parts; causes instability.

INCLUSION A principle of semantic relationship used in taxonomies wherein the terms at higher levels can be viewed as including those at lower levels; for example, "plant" includes "trees," "bushes," "flowers," and so forth.

INFANT MORTALITY RATE The number of deaths per 1000 persons under one year of age.

INFANTICIDE The intentional killing of infants.

INFORMATION A measure of uncertainty about the occurrence of an event or process.

INTERFERENCE COMPETITION Competing populations directly interfere with access to limited needs and wants.

J-SHAPED GROWTH PATTERN "Boom-bust" growth.

KAIKO RITUAL Tsembaga Maring (New Guinea) ritual playing an important role in information processing and feedback control.

KEY (LINGUISTICS) A network based on a series of hierarchically arranged attributes by means of which the identity of a form in question may be determined, for example, a key to identify species of coniferous trees.

KINETIC ENERGY A form of mechanical energy associated with the motion of a body.

KURU An infectious disease of the brain found among the Fore of highland New Guinea and transmitted by cannibalism.

KWASHIORKOR Protein deficiency disease prevalent in very young children in populations with inadequate protein intake.

LAND TENURE The rules regulating how land is owned and used.

LARGE CALORIE OR KILOGRAM CALORIE 1000 calories.

LAW OF LEAST EFFORT States that farmers will always make decisions to minimize their work load.

LAW OF THE MINIMUM Population growth, according to this law, is restricted by materials that are in the shortest supply relative to their need.

LEXEME A word, expression, or phrase that constitutes a separate unit whose meaning cannot be inferred from any other aspect of the language, for example, "jack-in-the-pulpit."

LIFE TABLE More accurately called death table, a table giving the chances of an individual's surviving from one age group to the next.

LIMITING FACTORS Those factors in the environment that limit population growth.

LINEAGE A kin group that traces descent from a common, known ancestor.

LITHOSPHERE The solid crust of the earth.

LOGISTIC GROWTH A model of population growth resembling an "S"; the population grows slowly at first, then exponentially, and finally levels off at the "carrying capacity."

MALTHUSIAN GROWTH A model of population growth in which growth is assumed to be controlled by limited resources.

MANA A dangerous, sacred power making a person taboo; a common belief in Polynesia.

MARITAL DISTANCE The distance between the birthplaces of marriage partners.

MARITAL MOVEMENT Geographical movement involved in the selection of mates.

MARKET A mode of exchange of goods and services among persons who are not socially related; impersonal exchange.

MATERIAL CYCLE A circular pathway followed by inorganic substances in an ecological system.

MATRILINEAL Descent from mother to daughter.

MAULIQTUQ A method of hunting seals by closely watching their breathing holes in ice.

MAYORDOMIA A Mesoamerican institution in which wealth is exchanged for prestige.

MEAN MATRIMONIAL RADIUS The average distance between the birthplaces of marriage partners and where they were married.

MECHANICAL ENERGY Energy capable of doing work.

MESOLITHIC A time period between the end of the Pleistocene and the beginning of effective farming; sometimes used to refer to post-Pleistocene adaptations of foragers in Western Europe and the Near East.

MODELS Simplified representations of the real world, using pictorial, verbal, or mathematical symbols.

MORTALITY CURVE A graphical representation of the relationship between the probability of dying and individual age.

MORTALITY RATE The number of deaths occurring within a given group over a given period of time.

MOUSTERIAN Archaeological culture associated with Neanderthals.

MUTUALISM Interaction between populations in which both are dependent upon each other; commonly referred to as "symbiosis" in anthropology.

NEANDERTHAL Early representatives of Homo sapiens, living about 100,000 to 40,000 years ago in the Old World.

NET PRIMARY PRODUCTIVITY Energy fixed by photosynthesis minus respiration losses.

NET REPRODUCTIVE RATE The number of female children that replace each female of the previous generation.

NEW YAM FESTIVAL West African ritual important in making decisions about when to start the yam harvest each year.

NONDARWINIAN EVOLUTION Evolutionary change by random processes.

NUCLEAR ENERGY Energy released from the nucleus of atoms.

NUMAYM Localized, bilateral kinship group among the Kwakiutl of the American Northwest Coast.

NUTRIENT CYCLE The circular pathway taken by materials essential to life.

OMNIVORE An organism that feeds upon both producers and consumers.

OPPORTUNISM The concept that the direction of evolutionary change is determined by the prior characteristics of the system, not by the invention of new characteristics.

OPPORTUNISTIC ORGANISMS Organisms that, because of physiological and behavioral flexibility, are able to rapidly undergo change from one life style to another and from one habitat to another.

ORDINATION The arrangement of ecological populations along environmental gradients.

PALEOECOLOGY The study of past ecological relationships from fossil and other material remains.

PALEOLITHIC Archaeological cultures falling within the time period from the earliest evidence of human remains until the end of the Pleistocene.

PALYNOLOGY The study of ancient environments from fossilized plant pollen.

PARADIGM (LINGUISTICS) A set of related terms sharing at least one attribute in common and differing from each other by at least one attribute.

PARASITISM Interaction between populations in which one feeds upon the other but, unlike predation, does not directly result in the prey's death.

PARENT/OFFSPRING MOVEMENT Total geographical movement involved in the selection of mates and reproduction.

PATRILINEAL Descent from father to son.

PHOTOSYNTHESIS The conversion of electromagnetic energy from the sun into chemical energy in the form of glucose; photosynthesis takes place mostly in plants and is necessary to sustain life on earth.

PHRATRY A social group of two or more lineages or clans, usually unilineal and exogamous.

PHYSICO-THEOLOGY The Medieval European concept that the earth is an "orderly, well-planned place," somewhat like a clock mechanism and controlled by divine purpose.

PIBLOKTOQ "Arctic hysteria" common among the Polar Eskimos and possibly caused by calcium deficiency.

PLASTICITY An irreversible physiological response to environmental stress.

PLEISTOCENE A geologic epoch, lasting from about 3,000,000 to 10,000 years ago, during which humans became the dominant creature.

POPULATION A group of individuals who cooperate to facilitate adaptation at some level, usually defined by a marriage network.

POPULATION DENSITY The number of individuals per unit area.

POPULATION ECOLOGY The study of the relationships between a population and its environment.

POPULATION PRESSURE A rather vague notion of what happens when population size approaches the carrying capacity.

POPULATION PYRAMID An ecological model expressing the relationship between population size and feeding habits ("trophic level").

POSSIBILISM The idea that environment limits but does not cause human behavior or biology.

POST-MARITAL DISTANCE The distance between the birthplaces of marriage partners and the birthplace of their last born offspring.

POST-MARITAL MOVEMENT Geographical movement determining where reproduction takes place after marriage.

POTENTIAL ENERGY A form of mechanical energy associated with a "standing" body and capable of being converted to kinetic energy.

POTLATCH An American Northwest Coast ritual in which valuables are given away or destroyed in exchange for prestige.

PREDATION Interaction between populations in which one population feeds upon the other, resulting in an energy transfer from one trophic level to another.

PRIMARY CONSUMER An organism that feeds directly upon producers, for example, herbivores that feed upon green plants.

PRIMARY PRODUCTIVITY The rate at which radiant energy is converted to glucose through photosynthesis.

PRINCIPLE OF UNIFORMITY The present is a clue to the past.

PRODUCERS Organisms capable of photosynthesis, including green plants and protista.

PROGRESSIVE EVOLUTION Evolutionary change in the direction of greater adaptability.

PULSE OR PULSATION A sudden spurt of energy or materials into a system, followed by a drastic reduction in its intensity or quantity.

RADIATION The direct movement of energy waves between objects.

RAMAGE A group of kin related to each other by descent from a common ancestor and ranked by distance from the main line of descent.

RANK SOCIETY A social organization marked by the hierarchical arrangement of status.

RATE OF NATURAL INCREASE The growth rate of a population, given by the difference between crude fertility rate and crude mortality rate, expressed as a percentage.

RECIPROCAL CAUSALITY Circular interaction between two things.

RECIPROCITY A mode of exchange of goods and services among social equals.

REDISTRIBUTION A mode of exchange of goods and services in a society with ranked forms of status.

RESILIENCE Processes that act to keep an ecological system from self-destructing, that is, to assure persistence even in the face of fluctuations.

RESPIRATION The exact opposite of photosynthesis, the breakdown of glucose by oxidation into carbon dioxide, water, and heat energy.

RITUAL Highly conventionalized, symbolic social behavior, possibly related to information processing.

ROLE A social function.

RUMBIN Sacred tree of the Tsembaga Maring (New Guinea) important in the *kaiko* ritual.

SAMPLING A method of using a small part of a population to infer population characteristics.

SAPROTROPHS Organisms that feed directly upon dead matter and are the first link in detritus chains.

SCRAMBLING COMPETITION Competing populations consume limited needs or wants with directly interfering with each other.

SECOND LAW OF THERMODYNAMICS The Law of Entropy, states that in a spontaneous process energy is continually degraded from a more-organized to a less-organized form.

SECONDARY CONSUMERS Organisms that feed upon primary consumers, for example, carnivores that feed upon herbivores.

SEDIMENTARY OR NONGASEOUS CYCLES The circular pathways taken by materials that do not exist at any time in a gaseous state, for example, phosphorus.

SEGMENTARY LINEAGE Social organization in which adjacent localized lineages join forces in opposition to more distant lineages.

SEGREGATE Any terminologically distinguished object or grouping of objects in a culture. Segregates form the elements of an ethnoscientific domain. (See Domain.)

SELECTION Genetic change through differential reproduction.

SELECTION COEFFICIENT An index of the intensity of selection.

SEMANTICS The study of meaning in languages and cultures.

SETTLEMENT PATTERN The distribution of human settlements in relationship to the landscape and to each other.

SEX RATIO A measure of a population's sex structure, given by dividing the number of females into the number of males and multiplying by 100.

SHIFTING FARMING A farming method in which fields are cleared from forest, farmed for a short time, and then allowed to go back into forest for rejuvenation.

SOCIAL ENVIRONMENT Relationships with other human groups.

SOCIAL SPACE A scale based on the intensity of social interaction between individuals or groups.

SOCIAL STRATIFICATION The hierarchical arrangement of status within a society by groups rather than by individuals.

SOUND ENERGY Energy in the form of waves that cannot be radiated in a vacuum but must have a medium such as air or water.

SPECIES The large group of organisms that can freely interbreed with each other and produce viable offspring.

SPECIFIC MORTALITY RATE The number of deaths per 1000 persons of a given age or sex during a calendar year.

STABILITY The degree to which the behavior of a system is maintained with given limits.

STABLE POPULATION A population in which age specific fertility and mortality rates are changing at the same rate. (See Static population.)

STABILIZING OR NORMALIZING SELECTION Selection that removes novel genotypes that are not advantageous.

STANDING CROP The amount of biomass present at any given time.

STATIC POPULATION A population in which age specific fertility and mortality rates are constant. (See Stable population.)

STATUS A social position.

STOCHASTIC Random.

STRATIGRAPHIC ASSOCIATION Sealed together in the same geological stratum.

SUBSIDIZED AGRICULTURE Agriculture dependent upon fossil fuels and other "outside" sources of energy.

SURVIVORSHIP CURVE A graphical representation of the relationship between the probability of surviving and individual age.

SYMBIOSIS Beneficial interrelationships such that one cannot get along without the other. (See Mutualism.)

SYNECOLOGY An approach to ecological study focusing upon the broad interrelationships among organisms.

SYSTEM A set of objects and their relationships.

SYSTEMS ECOLOGY The study of systemic relationships between living and nonliving things.

TAPHONOMY The study of the laws of burial.

TAXONOMY A set of terms and their referents arranged hierarchically from general to specific, for example, a taxonomy of flowering plants. (See Inclusion and Contrast.)

TERRITORIALITY Exclusion behavior toward habitat.

THERMODYNAMICS That branch of science concerned with energy and its transformation from one form into another.

THREE AGE SYSTEM A method of artifact classification devised in the nineteenth century by C. Thompson in Denmark. The system grouped artifacts into stone, bronze, and iron categories.

TIME MINIMIZER A feeding strategy in which the organism minimizes the amount of time devoted to feeding.

TIME-MOTION STUDY Documentation of energy expenditure in human populations by observing and recording work.

TRANSHUMANCE Seasonal movement from one ecological zone to another.

TRIBE A social organization made up of interrelated bands tied together by integrating mechanisms, such as secret societies, age grades, and other ritual institutions.

TROPHIC LEVEL A group made up of organisms with the same feeding habits, for example, producers, herbivores, or carnivores.

TRYPANOSOMIASIS Sleeping sickness, a disease caused by the trypanosome protozoa and carried by the tsetse fly.

UNILINEAL DESCENT GROUP A kin group that traces descent through either the male (patrilineal) or female (matrilineal) line.

VARZEA The floodplain of the Amazon River.

INDEX